ROSES
TO
RODEOS

Glorianne Weigand

Glorianne Weigand

ISBN 0-9644141-2-0

Publisher:
101 RANCH
Star Rt. 2 Box 31
Adin, California 96006

Printed by • Maverick Publications, Inc.
P.O. Box 5007 • Bend, Oregon 97708

Table of Contents

CHAPTER 1   The Fair Beginning   9

CHAPTER 2   The Directors, Loyalty Unsurpassed   27

CHAPTER 3   The Men Who Made it Happen   37

CHAPTER 4   The Women in All Their Glory   53

CHAPTER 5   Through the Years   63

CHAPTER 6   Snake Eyes   93

CHAPTER 7   Events of the Fair   95

CHAPTER 8   Livestock, the Hustle and Bustle of the Fair   105

CHAPTER 9   Action Packed Rodeos   121

CHAPTER 10   Get Ready for the Parade   135

CHAPTER 11   The Themes Contest   139

CHAPTER 12   The Steam Tractor   141

Foreword

To those that have been involved with the Inter-Mountain Fair from the beginning, may you enjoy this account of those events described in this book. Or even to those that may have only been slightly involved, or not at all.

This book is meant to commemorate the many years of the fair and give a history of that occasion.

In most puzzles the pieces are there and all you have to do is arrange them for a finished product. The pieces to this puzzle of seventy-seven years of the fair were not all there. Many years there was nothing to gain information from therefore the puzzle is incomplete. Some years just a small newspaper article, or a rodeo judging sheet scribbled on a piece of cardboard. Premium books, audits, handwritten notes or old newspapers were the source of information.

Invaluable people that shared their information with me, Albert Albaugh, George Ingram, Al Schofield, Shirley McArthur, Evelyn Eastman, Ione Strickland and Lottie Carpenter were among some that gave me information to print in this book. To them, Thank You for sharing some of the puzzle pieces.

To all of those that shared photographs, Valerie Lakey for the cover photo, Tammy Kofford for many of her excellent photos. Chuck McCully, Dorotha Kramer, Barbara Shaffer and Al Schofield for many of the old time Rodeo pictures, also Hack Lambert's family, and the boxes of old photos stored at the fairground. To those I have not mentioned, "Thank You". Without the help of all of you this book would not have been possible.

Glorianne Weigand

Dennis Hoffman

The Inter-Mountain Fair Board of Directors

Dedication

This book is dedicated to George Ingram who for over forty years gave of himself for the betterment of the fair.

Also this is dedicated to all the board members past and present that so freely gave their time for something they believed in.

To all the volunteers, that without you this fair would not be.

It may be said that this book is dedicated to all who love the fair.

The Fair Beginning

"REGULAR WILD WEST SHOW". The Inter-Mountain Fair was started in 1919. In the *Fall River Tidings* newspaper a large advertisement encouraged all readers to come to the "Fall River Valley Community Fair". October 2-3-4, 1919. $800.00 in cash prizes, Best Rider, 1st money, $50.00, Second Best, 2nd money $25.00, Third Best, 3rd money, $15.00.

Cash prizes for Best Hogs, Bulls, Horses, Mules and Cattle.

The fair consisted of a rodeo, being held in McArthur's corral and exhibits were shown in the forestry building, later known as the McArthur Grange hall and the Rose Barn.

William Albaugh, Roderick McArthur and James Day were the first fair committee.

The paper reported that the crowd assembled to do honor to our center was far beyond the capacity of both McArthur and Fall River Mills to accommodate. Arrangements had been made out of town to accommodate guests in farm houses within easy motoring distance. It was sad to realize that some people took advantage of our guests and charged outlandish prices.

Many expressions of admiration were heard for the decorations and arrangement of exhibits.

A variety of fruits and vegetables were shown by the various farmers. Apples were the most frequent and varied, but pears, peaches, plums, strawberries, and raspberries covered the exhibit tables.

Several coops of fowls added to the interest of the exhibition. Mrs. Wm. Wendt's beautiful Black Minorkas divided attention with her husband's attractive exhibit of big red apples and Hungarian prunes.

Mrs. D.C Bosworth and Mrs. J.H. Creighton along with Mrs. Katie Callison were honored for their prize fowls.

The Fall River and Glenburn flour mills had good exhibits of flours and cereals, and the cheese factory and meat packing company showed what they produced and provided samples for visitors to taste.

The Girls Canning Clubs of Hat Creek and Glenburn had a very creditable display of canned fruits and vegetables. Vera Opdyke of Hat Creek, and Ada Reynolds of Glenburn took first prizes and Doris Ratledge of Hat Creek and Evelyn Hollenbeak of Glenburn took second.

A place of great interest to the ladies was the room upstairs where the "Better Babies Contest", was staged. Dr. Dozier and his assistant conducted the clinic and measured, weighed and otherwise tested the many babies presented before them by their more or less doting parents. The one hundred per cent baby was Walter Merton Callison, son of Mr. and Mrs. Merton Callison of Glenburn with Bob Bruce son of Mr. and Mrs. Joe Bruce running a very close second. Mr. and Mrs. Archie Hollenbeaks small son Marvin ran a close third.

The airplane flights proved a great attraction as it was such a novelty to nearly everyone present. The sight of the great "Dragon Fly" scuttling along the ground before rising into the air was great fun for the children and the Indians, who gazed open eyed and open mouthed at the spectacle and acquired added respect for the white man's genius. One old fellow remarked; "Some day white man fly 'way up—talk to God!"

The riding and bronco busting attracted an interested crowd. We all like to see our fellows "riding for a fall", and in most cases were not disappointed. Miss Annie Ingle, however, kept her seat until her mount lay down, and she stepped off. Some of the others treated us to a fine exhibition of ground and lofty tumbling.

The fair and rodeo were operated on a voluntary basis. The rodeo receipts were used to pay awards on the exhibits.

When the McArthur Grange was organized, it's members helped build an exhibit building which they used for their meetings.

The Hollenbeak family was a part of the fair from the very beginning. Evelyn was eight years old when the first fair was held in 1919. She can remember planning on what you were going to wear to the fair. You wore your best clothes and it was the social event of the year. The rodeo was just out in the open corrals

1923 Fall River
Grange Booth in
the Rose Barn.

McArthur Grange
Booth in the Rose
Barn 1923.

Walter Callison, winner of the
most beautiful baby contest, 1919.

Bob Bruce, 2nd runner up of the
most beautiful baby contest, 1919.

Annie Ingle waiting to mount her blindfolded bronc.
Roderick McArthur waiting to help and John Ingram snubbing the horse.

McCarty at the 1919 rodeo.

Rodriques, 2nd best ride 1919.

on the McArthur ranch. The adults would sit on the fences and the children would peek through the fence boards to watch the excitement.

The September 26, 1919 issue of the *Fall River Tidings* headlines reads "BIG CROWD COMING TO OUR FAIR, OCT. 2-3-4." This big event, the first Farm Bureau Fair to be held in this section of the country. The first fair called the Fall River Valley Fair had to be postponed one week because of heavy rain and was held October 9,10 and 11. It was postponed mainly because an airplane from Red Bluff that was to be on exhibit and piloted by a man named McHenry could not make it.

The Finance Committee, Roderick McArthur, Jim Day, C.C. Hollenbeak and W.H. Stevens, while they are well pleased with the very liberal donations received, feel that every farmer in the valley should cheerfully donate at least $2.50 without being solicited for it, as his share, to help the good cause. Even though the committee had collected $800.00, the more money collected, the greater will be the numbers of cash prizes to be offered.

The following well liked men of the valley have been selected to be floor managers for the Grand Ball to be held at Fall River Mills. Glen Fitzwater, Gordon James, B.F. Gassaway, Elmer Erickson and Ralph Bidwell. Dance tickets will be $1.50. Supper extra.

Mrs. George Reynolds, Mrs. Freeman Post and Mrs. W.C. Selvester, the committee appointed to arrange for the Home Department exhibits desire to inform the ladies of the valley that $150.00 in prizes will be awarded for their exhibits. Divided into three divisions according to age groups.

The livestock was judged by W.M. Carruthers. Those exhibiting were Roy Cessna, Jim Day, Minor Michelson, Pearl Bosworth, W.C. Selvester, and H.S. Johns Poland China Hogs.

The Baum Ranch exhibited Shorthorns. Jim Day, W.J. Albaugh, C.C. Hollenbeck, Joe Bruce, W.C. Selvester exhibited Holsteins. R. McArthur had a fine exhibit of horses and mules, while Virgil Vineyard, Tom Vestal, and H.C. Johns exhibited draft horses.

Percy Creighton exhibited a Hereford bull.

N. Reynolds of Hat Creek had an excellent ranch exhibit. Hugh Watson had an exhibit of cheese. A meat exhibit was displayed by the Fall River Packing Co. The flour mills from Glenburn and Fall River Mills exhibited flour.

Embroidering and fancy work were judged by Mrs. C.G. Scammon and Miss Esther Steinbeck, (sister of the famous author John Steinbeck).

The judges of Agriculture exhibits were Charles Lemm, and Ben Stroup.

The rodeo was an exciting event and was well attended. George Farmer, a local cowboy won first and Dennis Rodriques of Red Bluff was second.

Aub Burton was the rodeo boss. Riders at the first fair were, George Farmer, Roy Farmer, Bill Allen, Dennis Rodriques, George Campbell, Ed Johnson, Cale Guthrie, Roy McClellan, Sie Elliott, Montana Red, John Gerig, and Miss Annie Ingle.

George Farmer rode the Dixie Valley sorrel mare. Ed Johnson drew the gray Beldon horse and was bucked off. George Campbell also was bucked off of a brown McArthur mare. In the finals, Roy Farmer drew the Beldon horse and was bucked off. Dennis Rodriques drew a little bay horse, owned by little Bill Hollenbeck, and he qualified. George Farmer also qualified on a McArthur horse. They then took up a collection and George rode the Beldon horse. Bill Allen and Dennis Rodriques rode mules bare-back and Bill won this event.

Every rancher that had a rank horse was happy to bring them to the rodeo to see if they could be ridden. Most of the bucking string was furnished by George Campbell, George Farmer and Billy Walker. This group of men also had a spotted pony that did a few tricks to entertain the children, one of those children Lottie Campbell,(now Carpenter who is 90 years old) can still remember the funny antics of the pony.

Most families brought their picnic lunches, but some did not have this foresight and all the excitement caused a lot of people to go hungry.

The October 1, 1920 *Fall River Tidings* headlines read, "SECOND ANNUAL INTER-MOUNTAIN FAIR WAS GRAND SUCCESS DESPITE STORMY WEATHER." Wonderful Display of Farm Products—Good Riding—Everyone pleased.

Held on September 23, 24 and 25, 1920 the second Annual INTER-MOUNTAIN Valley Fair came to a very successful close last Saturday evening, after the last big dance was staged in Fall River Mills.

Nearly two thousand people were in McArthur on Friday and Saturday viewing the exhibits, demonstrations and entertainments. It is estimated that fully one thousand Indians attended the fair, and fully that number of white persons were present the last two days.

The Farm Center baseball team walloped Bieber in a 14 to 0 victory. That gave the Farm Center the right to play Adin on Saturday for the championship of the INTER-MOUNTAIN valleys.

At the rodeo Thursday several good riders were forced "to bite the dust" or be disqualified for hunting too much leather with their hands. Plenty of bad horses

1919 Bronc riding by Annie Ingle.

George Farmer riding high.

Cale Guthrie.

Ed Johnson, one of the Indian riders.

Montana Red.

George Farmer, making a dust.

Jess Sears leaving Flying Annie, 1925.

Cale Guthrie in one of the first rodeos.

kept the nervy contestants busy to stay up where they could look over the amused stand of spectators.

Several Indian bronc riders started riding the rodeo circuit. Roy McClellan, Jim Reeves, Pinky Burns and Yakima Cannutt.

The lady bronc rider Annie Ingle was making quite a name for herself. She was quite a hand. Even though a little bitty thing, she was as tough as nails. Annie was a half blood Wintu Indian that was raised in the Pit River Canyon. She started riding when she was very young to check the families cattle. Her nephew Al Schofield born in 1906 can still recall watching his famous Aunt ride. Annie rode in the first McArthur Fairs and Rowdys as she called rodeos in Idaho, Pendleton, Oregon and Cheyenne, Wyoming. While touring the rodeo circuit she met the man she would marry, Bob Studnick. A world champion bronc rider. His brother Frank Studnick was three times world champion and they both rode at the McArthur Rodeo. In 1924 Annie, Bob and Frank traveled to London, England with a bunch of bronc riders to put on rodeos. Annie was at first billed as Shotgun Annie, then later referred to as Mrs. Bronco Bob.

When Al Schofield rode a bareback bronc in the 1925 McArthur Rodeo he can remember the top hands that day were, Montana Red, Perry Ivory, Joe Moss, Cale Guthrie, John Gerig, Sie Elliott, Marshall Flowers, Howard Teglan, George Farmer and Annie Ingle.

Annie rode broncs and relay races for many years and rode with the best of them.

Plenty of hard mules and bulls were topped off and never once was a rider "busted" here. Exhibition riding by George Farmer and Joe Chico gave them much praise and admiration from the crowd by their wonderful exhibitions.

Livestock Day came on Friday and there were many animals and coops of fancy chickens shown. Most interest was shown the Calf Club boys, who had their Shorthorn and Holstein calves there to make a bid for the prizes offered. The late Senator Rush offered a large silver trophy cup for the best beef type Shorthorn calf in the boys club. The cup must be won two consecutive years by the same boy in order for him to keep it as his trophy. First prize along with the cup was $10.00.

Marvin Dimick of Glenburn won the big cup and the first place prize, as his Shorthorn heifer calf was judged the best of the lot. Edwin Albaugh with his heifer calf won second money of $10.00, and Byron Dimick third money of $5.00 with his bull calf.

The Girls Canning Club had the north end of the hall completely filled with a wonderful array of fruits, vegetables and meat packed in shiny jars by the cold pack process.

Miss Evelyn Hollenbeak won the canning club contest and exhibit, as she had over fifty cans of different edibles neatly packed away. Miss Hollenbeak, (10 years old), surely does deserve much credit for the amount of work and the skillful way she put up her cans. With first prize in the contest she was awarded a silver cup, also donated by the Senator Rush. She must win the cup two years in a row to be able to keep it, (which she did). She also won the chance to attend the Boys and Girls Conference at the University Farm at Davis. (Evelyn Hollenbeak Eastman still has the cup displayed in her home with pride.)

Miss Hollenbeak and Miss Vestal demonstrated very precisely every step taken in the cold packing of fruits and vegetables. These two little ladies talked and worked before a large crowd of women and impressed them with their knowledge of canning. Miss Hollenbeak also canned a variety of Indian foods, such as porcupine, acorns and apaw roots to add to her bounty.

Anderson Valley had a fine exhibit at this fair and it consisted of projects that could not be grown in Fall River Valley.

Agriculture expert, A.E. Gray, Dept. of Agriculture State of California said the products grown in the Fall River country surpass those from rich valleys. It was a well arranged and handled fair without a hitch. The people of the Fall River section deserve credit. Then 7J ranch from Hat Creek, owned by Mr. Reynolds had a fine exhibit which comprised of everything that could be raised on 160 acres.

Many visited the display of the Ever-Ready Aluminum cook ware, Monogram Oil Products, the Samson Tractor, G.M.C. Trucks and the Studebaker car.

Mrs. Rose and Mrs. Vernon Selvester were more than busy at their restaurants serving meals to the hungry folks. The ladies of Big Valley had a sandwich stand outside the grounds. Thus there was plenty to eat for all, and no one went hungry this year, as was the case last year.

The only disappointment on the program was the aeroplane, which did not show up, as was hoped, due solely to the bad weather. Storm and winds caused low clouds, and it was impossible for the plane to arrive from Willows.

It is hoped that the fair will be staged at least two weeks earlier next year so good weather can be expected.

And now we must look forward to 1921 to our Third Annual Inter-Mountain Fair and hope it is bigger and even better, (if that is possible).

1922, Mrs. Broncho Bob, (Annie Ingle) on Blue Dog.

Annie Ingle and Bob Studinick wedding picture.
He was world champion bronc rider and they both rode at McArthur.

Annie Ingle on the burro with some of her cowgirl friends.

Bob and Annie (Ingle) Studinick.

Skeeter Bill.

Evelyn Hollenbeaks canning display with her winning trophy at the top.

Scott McArthur had donated seventeen acres of land to be used for the fair. September the 3rd and 4th were selected as work days when all farm centers would participate in fencing this area and constructing a grandstand and bucking chutes. Lumber was hauled from the Horr sawmill in Glenburn for the grandstand, corrals and fences. The hauling had to be done with team and wagons and Roderick McArthur, Glen Fitzwater, Willis and Rube Albaugh were the teamsters. They had to make two trips a day with their teams and wagons.

Aub Burton was still Rodeo boss and some of the contestants that year were, John Gerig, little Joe Pacheco, Willie Fulsome, Roy Sutton, Ned McGrue and George Farmer. John Gerig drew the Robert's horse and was bucked off. The first prize in bronc riding was $150.00 and it was announced that the ride John made was worth $150,000.00. Joe rode Two Step and Roy qualified on Bald Hornet. Willie drew the Robert's horse the second day and was bucked off. Again they took up a collection for George Farmer to do an exhibition ride.

The third fair in 1921 opened with a baseball game. A band from Westwood played at the game. This was the first year for horse racing, and the first year livestock was exhibited.

In 1921 the fair committee reported the grounds were ready for the Fair. Additional bleachers and stock corrals have been built. committees were appointed by the director and chairman of the Farm Home department to take care of the various parts of the work as follows.

Arrangements in hall; W.E. Agee, U.L. Walker, J.C. Stanley, N. Reynolds, Mrs. C.C. Hollenbeak, Mrs. M.D. Nicholson, Mrs. D.C. Bosworth, Mrs. Harry St.John.

Clothing exhibit; Mrs. Oscar Hill, Mrs. Jim Bowman.

Food Conservation exhibit; Miss Lois Whipple.

Fancy work exhibit; Mrs. Jack Creighton.

Home Sanitation exhibit; Miss Anna Albaugh.

Hot school lunch exhibit; Mrs. Virgil Vinyard.

Home department prize committee; Mrs. H.M. Rice, Mrs. W.J. Albaugh, Mrs. Clarence Ayers.

Livestock prize committee; Parker Talbot, C.C. Hollenbeak, Perry Opdyke, Joe Bruce.

Livestock manager; J.R. Day.

Ticket and gate committee; H.M. Rice, M.D. Nicholson, Jack Ingram, Billy Ingram, Jim Bowman, Jesse Hill.

Dance Committee; Elbert Lee, John Ingram, Reuben Albaugh, Oscar Hill, Billy Ingram.

Concessions committee; J.A. Dawson, Pearl Bosworth, T.J. Dunlap.

Messrs. Reber and Woodmansee were appointed to look after getting an exhibit from Big Valley.

Many Indians from Dixie Valley, Big Valley, Hat Creek and Fall River Valley attended the Fair. There were two Indian camps one behind the Grange Hall and the other over near the fair grounds along the McArthur canal. The Indians would gamble with their stick or grass games all night long. Their chanting to be heard all through the night.

In their game of gambling they would have several on their team and the teams would sit facing each other. They would have sticks or rocks that they called "bones", that they would pass from one to another. One of the items would be marked with a black line or mark and that was the special bone. When the team with the special bone finished their chant and passing their bones from one to another they would put their hands down and stop at once. The opposite team would have to guess who had the special bone. If they won the guess they would win the money in the middle of the group or the sticks, whatever they were playing for. The magic bone would be passed to the opposite team and the game would go on with them passing the bone and chanting. If the guess was wrong, more money or sticks were added to the pot and you kept the magic bone until the opponents guessed right.

In 1922 the fourth annual fair was held. W.J. Albaugh won the dairy cow production contest and the Adin High School won the Dairy Cow judging. Leland Anders was the high individual in the judging contest. Fall River Farm Center won the Agriculture Exhibit over the Hat Creek Farm Center.

Bucking horses were starting to be contracted and Bill Simms from Oregon brought the horses that year. W.J. Albaugh brought Black Bart for one of the cowboys to ride. McKinley Jackson an Indian from Alturas rode him, but did not qualify.

Relay races were becoming popular and Reuben Albaugh and Jim Day teamed up for the relay races and they won all three days. Reuben had a favorite horse Zane Grey and he was hard to beat. Jimmie George, an Indian from Big Valley had a relay string that came in second. Rube raced his Zane Grey horse in the half mile race beating Cecil Day on Joe Sawyers mare and Eldon Bernard on Vestal's sorrel mare. This was a big upset and Rube won the prize money of $62.00. The local Blacksmith at Pittville, Press Fine, made a special set of race shoes for Zane Grey.

In 1924 the race event was marred when two horses with riders suffered bad spills. Harold Vestal's horse

Indian Gambling at the fair.

Indian stick games at the fair.

Ready for the race,
Bill Altizer on
Rube Albaugh's horse
Zane Grey.

Lil Bognuda riding a bronc.

rolled completely over his rider. Harold was riding Rubes Zane Grey when he and Jim Day's horse collided on the first turn of the track. No one was hurt.

Art Kinyon from Bieber won the race riding a bay horse.

The young men of the valley started taking their turns at organizing the fair. Rube Albaugh was elected President. Earl Cooley, Ag. teacher was Secretary and Bill Eldridge was Rodeo boss with Hardy Vestal as his right hand man. The young men needed to make some money to run the fair on. They went to Susanville and collected $100.00 from William Walker of the Red River Lumber Company. They then went to Redding to ask PG&E for some money, but they were turned down. The young men sponsored dances to make money for the fair.

A dispatch from Redding to the Sacramento Bee in 1921 says; Speaking of the fact that the Inter-Mountain Fair at McArthur will not get any money from the State's appropriation of $25,000.00 for the encouragement of county fair, Roderick McArthur of McArthur, one of the main promoters of the Intermountain Fair said: "Individually I am opposed to those gifts from the State. I believe they do more harm than good. Up our way we own the Inter-Mountain Fair. It is our institution, and we have the pride of being thrown on our own resources. We are not asking for gifts of charity.

We will have to take in $6,000.00 at our fair or we will be in the hole. We are not afraid. Our fair last fall exceeded all expectations, and we are doing more this season. We have doubled the capacity of our grandstand and doubled the number of stalls for exhibiting stock. We have doubled the area of the fair grounds and have put in a race track three-quarters of a mile long.

The fair we had last fall had a wonderful effect on the farmers and stock growers. It has inspired them to improve their breed of stock and to improve and widen their field of cultivation.

N. Reynolds of Hat Creek had the largest and best exhibit of grains and vegetables. He showed to other farmers what could be done, and a good many farmers this season have taken a lesson from Reynolds."

"We don't need their money. We'll do it our way and we will be proud of our Fair."

One of the founding fathers,
Roderick McArthur.

One of the founding fathers,
W.J. Albaugh.

The Directors, Loyalty Unsurpassed

The founding fathers of the McArthur Fair in 1918 were Roderick McArthur, W.J. Albaugh and James R. Day.

These three ranchers felt there was a need for a fair and rodeo in the McArthur area. The first rodeo was held in McArthur's corrals and the first exhibits were displayed in George Rose's barn.

Fair Board Directors are appointed by the Shasta Board of Supervisors and serve a term of five years.

They are involved in making decisions concerning business activities and operation of the fair. Unless someone wants to retire or moves out of the area the same board is usually reappointed. There have only been fourteen members to date, including the three men who started it all, W.J. Albaugh, Roderick McArthur and James R. Day. A span of seventy-eight years these faithful members have done a magnificent job.

Willis Albaugh son of W.J. Albaugh was appointed

One of the founding fathers, James R. Day.

Fair Director, Willis Albaugh.

to the board of directors of the Inter-Mountain Fair in 1922 and retired in 1978. This was a term of service of fifty-six years for Willis. Willis was honored with the Blue Ribbon Award in 1969 a most deserving recipient. Willis Albaugh always went that extra mile for the fair and did his best to promote and beautify the grounds. He could be seen planting trees, pouring cement, fixing fence, working on the buildings helping with the livestock, or anything that needed to be done. Willis was there and served as President of the board for most of his term in office. The Inter-Mountain Fair was his pride and joy.

Asa Doty, a member of the board for 25 years retired so he could spend his winters in the warmer climate. Asa, a Hat Creek rancher, was a valuable member of the board. A lot of hard work was done while Asa was on the board, building an admirable fair grounds from the beginning.

Byron Hollenbeak was only on the board for one year when he was president of the board in 1941. He then turned his efforts to working with the rodeo. He headed the rodeo in 1928 and 1929. Byron would rather be on a horse than sitting at a board meeting.

Hugh Carpenter was appointed to the board in 1949. Hugh was a member of the ranching community of Fall River Valley. Hugh was also a member of a pioneering family that was very much involved with the fair and rodeo. Hughs' father-in-law James Day was one of the original three founders of the fair. Hugh sold his ranch and moved away in 1958. This left a vacancy on the board that was filled by George Brown Jr.

Willard Brown joined the Fair Association in 1956. The fair was growing and more board members were needed to make the decisions. Willard was born and raised in Fall River Valley and a owner of Sierra Market, he had a true interest in the welfare of the community and the progress of the fair. Serving twenty-three years of the fair board was a pleasure for Willard Brown.

Wallace Hilliard from Burney was appointed to the Fair Board in 1956 and served as President of the board in 1978. Known as Wally to his friends he was a valuable member of the board of directors and was honored with the Blue Ribbon Award in 1990. The Hilliard Insurance in Burney was where Wally put his

efforts when not working with the fair.

George Brown Jr. was appointed to the board in 1959. A heavy equipment operator and farmer he had a sincere interest in the fair. George left the area in 1972 leaving his position open to be filled by Bud Knoch.

Catherine Ryan took Asa Dotys' place and Catherine was on the board for nineteen years and resigned so she would have time to participate in other activities. A rancher from Hat Creek she was noted for having done a tremendous service to the cattle industry. Catherine was the recipient of the Blue Ribbon Award in 1979. This award is given yearly to a member of the community that has devoted their time and energy to the betterment of the fair. Many volunteers in the community have been honored by receiving this award. Catherine was on the board from 1965 to 1983.

Leland (Bud) Knoch joined the board in 1972 replacing George Brown Jr. when he moved away. Bud is a member of a pioneer family in the McArthur area and has grown up showing animals and being a part of the fair. He is in the ranching business with his family.

Albert Albaugh replaced his brother Willis on the Fair Board when Willis retired in 1978. Albert is the third member of his family to serve on the board and follows his fathers and his brothers footsteps. His father came to the valley as a young man and made his mark in the community as an outstanding member of the agriculture world.

Bob Thompson Jr. took Catherine Ryans place in 1983. Bob is a rancher and contractor from the Hat Creek Area and has been involved with the fair most of his life. Like other members of the board Bob's family are pioneers in the ranching business in the area.

In 1989 Skip Willmore was appointed to the board to fill the vacancy when Williard Brown passed away. Skip is the owner of the popular Clark Creek Lodge. Raised in Burney and graduating from Fall River High School before Burney had their own high school.

In 1990 Craig McArthur became a member of the board when Wally Hilliard resigned. It seems only fitting that Craig would fill this vacancy to follow in his grandfather Roderick McArthurs footsteps. Craig is a rancher and is preserving and making his home in the beautiful landmark home that Roderick McArthur built for his family.

Asa L. Doty

Fair Director, Hugh Carpenter.

Fair Director, Byron Hollenbeak.

Fair Director, Willard Brown.

Fair Director, Wallace Hilliard.

Fair Director, George Brown Jr.

Fair Director, Catherine Ryan.

Fair Director, Leland "Bud" Knoch.

Fair Director, Robert Thompson.

Fair Director, Albert Albaugh and Lem Earnest.

A Fair Board Meeting, Willard Brown, Willis Albaugh, Wally Hilliard, Catherine Ryan, George Brown Jr.

1952 picture of the McArthur Fairgrounds. Notice the lack of trees.

Back L. to R.:
Jesse Bequette and George Ingram, Front L. to R.: Mary Albaugh, Willis Albaugh and Wally Hilliard.

The Men Who Made It Happen

A MAN AMONG MEN

Parker Talbot

Parker Talbot joined up with the Agricultural Extension in the early 1900's, shortly after Extension became established in California. Parker was Shasta County's first farm advisor, and Shasta County was Parker's first big job. He moved to the county in

Parker Talbot, a Living Legend.

October of 1917, and was later transferred to San Luis Obispo County as county director. He retired from that post and from Agricultural Extension in 1949, after thirty two busy and constructive years. In both counties, Parker remains a living legend.

Farmers in Shasta County and the Inter-Mountain area can vouch that Parker has been good to, and for them.

The beginning of the Inter Mountain Fair is attributed to the leadership of Parker Talbot. Roderick McArthur, W.J. Albaugh and Jim Day were the first committee men, but under the strong leadership and influence of Parker Talbot. Parker cultivated both qualities of leadership and cooperation in the outstanding ranchers he attracted to the work of improving agricultural methods and rural living conditions.

A young farm advisor with a quick smile and keenly humorous eyes, he cast a warm spell over all who worked by his side. At the beginning of his career he earned $1,500.00 a year and drove a donated $500.00 Ford that he didn't know how to operate.

One of Parkers most outstanding accomplishment was the founding of two fairs. One at McArthur and one at Anderson. The Inter-Mountain Fair is still run by some of the sons and grandsons of the men associated with Parker Talbot in the original venture.

Parker is responsible for many good things in Shasta County, that include the building of the Farm Bureau, A mosquito abatement district, the elimination of malaria from the district, took the lead in helping dairies clean up TB and Bang's disease. Organized the livestock shipping association, organized the rural telephone and rural road system and organized the Production Credit Assoc. in 1933.

Parker wanted W.J. Albaugh to accompany him on a very important Farm Bureau Traveling Conference in 1920. He knew that if he merely wrote and asked Bill to go, the answer would probably be no. Rather that risk that, Parker decided to go in person to the Albaugh ranch at McArthur.

The month was February and the mountain road over Hatchet was impassable. But this didn't stop Parker. He boarded the train at Redding and rode it to Sisson (Mt. Shasta), then traveled by the McCloud railroad to McCloud were he stayed overnight. The next morning he rode the lumber company logging train to Bartle, then by four horse stage to Glenburn, (thirty-four miles). Jim Day met him with a two horse wagon, then it took two hours to travel the six miles to McArthur with snow and mud all the way. Only a few came to the Farm Bureau meeting that was scheduled. Parker announced that he made the trip with only one thought in mind, to make Bill Albaugh a delegate to

the Farm Bureau Conference. Bill said he couldn't be away from home so long as he couldn't get across Hatchet Mountain. Parker was grimly disappointed. Roderick McArthur leaned over and said to Parker, "Bill will be in Redding at the appointed time, I will see to it." McArthur kept his promise.

A LEADER, A TEACHER, A FRIEND

Jesse Bequette

Being a teacher of Agriculture in McArthur, Calif. can be quite an experience to say the least.

Jess Bequette took his job seriously in 1936 when he first came to the Fall River Valley. The men today that had him as their teacher so many years ago still talk fondly of him. He believed in teaching Agriculture to the fullest. He taught the boys to sharpen saws and axes, how to tie knots in ropes, how to work on equipment, build fences as well as care for and raise livestock. He was going to make ranchers and farmers out of them, and he did. He was the instigator that got so many of the young boys started in the registered cattle business or livestock of some sort in the Fall River Valley.

When you took on the job of Ag. Teacher at the local High School you automatically got the job of running the Inter-Mountain Fair. Jesse Bequette had his hands full with both jobs. But he had some young energetic students that worked hard and helped Jesse keep the fair going.

A letter Jesse Bequette wrote in 1978 sums up a lot of his experiences at Fall River Valley. Some excerpts of his letter are as follows.

My first visit to the fairgrounds in July 1936 was to find a small lot of perhaps five acres with a fallen down board fence on the East and South sides, a straw covered shed for its only livestock barn along the East, outside of the fence. A dilapidated open bleacher that had stood there over twenty years. All Rodeo chutes and corrals were in bad repair. Adjoining the grounds on the South side, a little West of the bleachers, was a building that belonged to the McArthur Grange and was used by the

Jesse Bequette.

Fair for its only exhibit building. An old fashioned two hole toilet stood by the rodeo corral and the Grange had the same kind close to its hall. The stock shed had room for about forty hogs and sheep. Cattle were tied to the fence posts out in the open at fair time.

Board members as I recall, were Asa Doty, Willis Albaugh, Hugh Carpenter and Leland Anders. Secretary, Frances Gassaway and Treasurer, Clarence Whipple. Willis managed the fair and Leland managed the Rodeo. I immediately accepted the responsibility of overseeing a local crew of volunteer men to set the old fence up. That turned out to be a job, but we got it done.

Labor Day and fair time arrived. About forty to fifty head of cattle, mostly Hereford, and a few hogs and sheep were entered. A real nice group of forty head of riding horses came for the halter classes. The exhibit hall was filled with garden, grain, hay and such. There were agriculture exhibits and some Grange and F.F.A. Booths.

Livestock judging time arrived, exhibitors were ready, and no judge. After an hour or so of delay, Willis asked me to do the honors. That I did. Well I remember not giving a blue or a red ribbon in the swine division. The animals were of such poor quality that I started with third placing and told why in giving reasons. I believe that judgement did a lot to improve quality in the swine of that valley in later years.

Judging the cattle and horses was easier, as their quality was reasonably good for Grade animals. I think the only registered animals on the grounds was one or two Stallions, owned by Jess Eldridge and Ward Kramer. Kramer came from Big Valley.

The next day started with a nice morning shower. But soon trouble developed among the show cattle. The cattle tied to the posts had eaten the wet alfalfa hay and were bloating. I soon had everyone walking the bloated animals with gags in their mouths to get them to belch the gas out. Soon Perry Opdyke who was leading the shows Champion Hereford heifer, yelled, "She can't move, she's dying!" I ran over opening my pocket knife and jabbed a hole in the proper spot in the heifer's left side, puncturing the paunch. A stream of air shot out that knocked my hat off and you could smell some distance away. Perry said, "Oh no, you have ruined her." In an hour or two she was O.K., and in a week, Perry could not even find her scar. Sticking, as it is called was our only chance of saving her. This was a method I had learned in Montana and it was used quite often up there.

About five hundred people jammed the rickety old bleacher and sat on the tops of the old wooden rodeo fence to watch the afternoon rodeo.

That evening I met with about thirty cattlemen in the High School Ag room to discuss my helping them purchase a carload of purebred bulls in Montana. They elected me to make the trip and pay up to $150.00 a head for the bulls delivered by rail to Bieber. I was allowed a $10.00 limit per head in addition for the other expenses. Mr. Lee and his son John and grandson Walter Callison made the trip to Montana with me. We had to choose 30 head out of a herd of about 300 in a large open corral. The cold wind blown prairie near Belt, Montana in November of 1936 is where we chose the bulls. When they were delivered to Bieber a number was written on each bulls horn and lots were drawn for the bull you would receive.

In 1937 the fair board invited Jesse to design and take charge of building a covered grandstand adjoining the old bleachers on the West end. Willis and Jesse drove to Sacramento where they presented the plan to A.E. Snider, Chief of the Horse racing fund for the State fairs. After much discussion $1,400.00 was authorized to construct the new grandstand for about 500 seating capacity. In 1936 our total funds, other than rodeo gate receipts, had been $500.00 for the fair contributed by Shasta County.

After returning to McArthur, I went to Mr. Hawkins, owner of the Hawkins Saw Mill in Dana, to order all the pine lumber for the Grandstand. He mentioned that he had not been paid for the lumber bought twenty-five years ago. I convinced him that I had the money to pay him. Arthur Dunlap offered to furnish the sheet metal roofing and hardware at his cost plus five per-cent. Wally Belden built the Grandstand, but we were $200.00 short, but after negotiation the State paid the rest of the tab.

The 1937 fair was larger with more livestock and exhibits. Water for all the livestock became a problem as the Road Department hauled it in a couple of tanks and were chastised by the county head of their department.

In 1938 the old straw shed had collapsed during the winter. A descent livestock barn for cattle, sheep and hogs was needed. A well was also desperately needed. Again Willis Albaugh and I went to Sacramento to visit Mr. Snider with the plans. After a serious discussion we were authorized $1,000.00 for the project plus $500.00 for premium money. The new additions ran over their expected cost by $250.00, and again Mr. Snider approved the state's payment.

The old grandstand was lowered by cutting of nine feet at the bottom and was made into a barn. That year the fair exhibits increased quite a bit and it took two

Bethel Brown and his Shorthorn heifer.

Floyd Bidwell. 1939.

Jesse Bequette's Ag students showing cattle, 1939.
Morris Doty, Carson Estes and John McArthur are the first three.

Jesse Bequette's Ag students, 1939. Asa Lakey and Joe Sawyer showing pigs.

Some of the bulls that Jesse Bequette brought from Montana in 1936. His ag boys L. to R. Harry Brown, Elmo Brown, Paul Loftin, Bethel Brown, David Shaffer, Lem Earnest, Ward Kinyon, Gordon Opdyke, Robert Shaffer, Joe Shepherd.

judges to do the livestock. One for the horses and the other did the rest of the livestock. Beef Cattlemen began asking for a Range Cattle division. But first, it would be necessary to construct proper facilities. In early 1939 Willis said the Fair Board wanted us to construct another barn, one for horses only. A new barn was built for $1,100.00 which included a water line to it and two outside toilets.

In 1940 another barn for cattle was planned and built with State money and about $700.00 awarded for premiums. In early June, Asa Doty and Jesse attended a Western Fairs Association meeting in Stockton for Northern California Fair Managers and Directors. Four large fairs in the state had been receiving the race track monies to run their fairs on and the smaller fairs were doing without. I was appointed spokesman for the smaller fairs. A couple of the larger fairs were receiving a half million dollars annually to run their fairs and ones like Inter-Mountain was getting by on peanuts. Before the meeting was over a motion had been made and seconded and a few red faces of the larger fair managers dropped when a motion passed that the small fairs should get their just amount. In a couple weeks a notice was received that the Inter Mountain Fair would receive $5,500.00 for the 1940 fair to use for premiums and a few other items.

Plans were drawn up and pens were built for the range cattle show. The state required a pen to consist of ten steers, or heifers, in calf, and yearling classes. With only a top premium of 1st, $35.00; 2nd, $25.00; 3rd, $15.00. A few pens were shown in 1941 and 1942 as it took nearly a year to build the facilities. Fred Schneider was contracted to cut cedar trees and split them into big, strong, long lasting posts, the rough pine boards were purchased at Hawkins sawmill and all the hardware came from Dunlap.

Interest was rapidly building among 4-H and F.F.A. members in exhibiting many breeds of pure-bred livestock. The added premium money brought increased entrees each year from 1940 on. Each year we paid out all the premium money permitted by the State. County commissioners stopped the $500 county tax money contribution in 1942. Up to 1943 we had over extended our State building allowance by $150 to $300 each year. Because of the war there was no fair in 1943. Then in October 1943 an Auditor arrived from the State Division of Finance to audit our books.

The auditor, Eric McLaughlin, wrote to our faithful secretary, Frances Gassaway, to have a long list of items in a box for him to pick up on a certain date to take to Sacramento. When he arrived in Fall River Mills at the Gassaway residence about 11 A.M. no one

was home, so he drove to McArthur to find me in a classroom. He told me his troubles and that he was very much upset. I told him I had not been informed of his request, but to go to the Fall River Mills Grammar School where Frances was teaching, that I was sure he would find her there. He found Frances and she gave him her house key and information that the box of materials were under her bed and she thought all items listed in his request were in it. There he found them. But upon examination with his check list found that the canceled checks were missing. Giving the key back to Frances she said, "Oh, Clarence Whipple, our Treasurer, has those". She gave the irritated Eric instructions on how to find the Whipple ranch. At this point Eric bawled Frances out for not following HIS instructions. (That was a mistake.) Then Frances told him to go to thunder, that he was being paid for his job, but every one of us local workers were donating our time, gas and etc.

At Whipple's ranch, Mrs. Whipple told Eric that Clarence was out across the river at the back corral loading a large truck with their steers for market. He drove across the bridge at the Big Spring, and turned into the field. Observing it muddy and slick, fearing he would get stuck he parked his State car, leaving a couple windows open and took off on foot for the big truck he could see about a quarter mile across the field. Finding Clarence unwilling to stop the loading process to wait on him, he got angry with Clarence. That was another mistake. Clarence told him to get out of the way and stay out of their way, when the cattle were gone he would get the canceled checks and ledgers. Eric waded, slipped and slid through the mud back to his car. There, to add fuel to his anger, he found several turkeys in his car, (from a farm across the road). He just finished cleaning turkey manure from inside his car when the cattle truck went by. Back at the house, when Clarence handed Eric the items he wanted, neither man was in a good frame of mind Clarence informed Eric his time had been volunteer and he would take no gaff from State punks on State payroll.

When McLaughlin's audit report came back it was our turn to be disturbed. The report showed the Fair had illegally paid out over $7,000.00 in state monies for premiums, and about $1,800.00 on building over our budgets during the past years that we had been getting Mr. Snider's approvals. The auditor disallowed Mr. Snider's approvals on out previous building over runs. This really shook the Fair Board. They asked me to go see Snider and find out what could be done since our treasury was broke.

Upon discussing the situation with Mr. Snider and

George Miller, they shrugged their shoulders over the $1,800.00. But the Premiums, we found had been paid on our range classes, horse classes, and several of the cattle and hogs in all the grade classes when entries had not shown they were sired by a purebred registered sire. If we could get a clearance on every entry we could dig up a sire registration number for, and swear by the owner's signed affidavit that he knew it to be that animal's sire, we would be in the clear.

The next few weeks I burned a lot of school pickup gas getting entry owners to sign registration affidavits. We were fortunate that I had typed records of each purebred bull, boar, and ram I had purchased and brought into the Valley. Jess Eldridge, Jack Metzger, Ward Kramer, Bill Lee, Leland Anders and Hugh Carpenter owned the registered stallions that had sired the horses that had been shown. In six or seven weeks I mailed a bundle of affidavits to Snider that canceled all but about $150.00 of the over $7,000.00. That was a terrific relief. Another discussion with Snider eased out minds about the $1,800.00, and the $150.00 remainder. He said, "Let it ride, and don't worry, we know it went into the Fair buildings, and you have done a good job building them".

In 1944 and 1945 we received about $3,500 in State funds. Each Fall audit reminded us of our former $1,950 in arrears we had been unable to do anything about. We continued to follow Mr. Snider's suggestion.

In 1944 Cattlemen complained of the small premiums for such large pens of cattle. Premiums of $35., $25., and $10. The matter was discussed with Snider. On paper I pointed out the shrink on a herd of cattle to select a pen and the cost of trucking to and from the Fair, a man's time to care for them during the three or four days. It all totaled up to over a cost of $100.00 per pen. That, in the heart of a real range country cattlemen should be encouraged to compete in their profession and at a premium large enough for the winner to break even. They agreed to allow us premiums for each of six classes of $125, $100, $80, $60, $40 and $20. That year, and the years following the McArthur Range Cattle Show was larger than that at the State Range Cattle exhibits at the Sacramento State Fair.

In 1944, the Rodeo committee gave up and said they had not made money, and refused to continue promoting the Rodeo. In desperation, the Fair Board debated what to do. By this time I was sponsor of a very strong Young Farmers Chapter. These were young men who had been in our McArthur F.F.A. and had become ineligible to be an F.F.A. member one year after graduation from High School.

John McArthur, Floyd Bidwell, Dick Norris, Asa Lakey, Andy Lakey, Carson Estes, Morris Doty and Buster Hawkins asked me if they could take over and put on the Rodeo. That they did, with the Fair Board's blessing. This automatically placed me in a very responsible seat at its head, since I was legally their advisor. The boys felt they could not afford to hire bucking horses and roping stock. They got a couple of cattlemen to donate roping stock. The boys made plans to take saddle horses in two or three large stock trucks out to the "Wild Horse Range" North East of McArthur, the day before the fair, round up wild horses and truck them in for bucking stock. I was to go with the eight boys, but something came up that morning to detain me. They went out and returned with 25 or 30 of the wildest horses I believe ever used in a rodeo.

Since management and promotion was under school management we set a rule of no liquor to be used on the Rodeo ground.

A pop company had ice cold pop everywhere to keep the cowboys thirst quenched.

The Young Farmers ran off some eighty bucking and roping events in ninety minutes. Never a dull minute kept the crowd waiting during the show. The only casualty was when a wild horse in a chute struck Dick Norris in the mouth, badly loosening several of his teeth.

That group of boys ran the rodeo the next year and were successful financially both years. But one year of using wild horses was enough, the second year they rented a Rodeo bucking string from Dick Hemstead. After the rodeo committee watched the boys have two successful years, they ask if they could take it back.

After the 1945 Fair, in about October, word came from our audit and it showed a clean slate. How, or why we will never know but we were very pleased. The next spring word came from the State division of Fairs and expositions that each of the fairs in California had been awarded $65,000.00 from the enormous Race Track Fund. Immediately the local Fair Board asked me to draw up a master plan for future expansion.

The Fairgrounds at that time consisted of the approximately 20 acres that had been donated by Scott McArthur in about 1920. It had been deeded to the High School. Negotiations got underway to purchase some land from PG&E and there were two blocks of the town of McArthur lying between the fairgrounds and the swamp. The master plan we set up moved the grandstand to the present location with new rodeo chutes and corrals, one or two exhibit halls, livestock and horse barns, master well and modern rest rooms

43

in the exhibit buildings and one outside with showers. By the time these had been approved in 1946 I was leaving the teaching job at McArthur. The Fair Board invited me to stay on and receive a salary from the Fair (the first salary ever paid by that Fair), at least until the new plan could be put into operation.

During the time that I spent at the McArthur Fair Grounds, (ten years and three months) I received a lot of pleasure in helping to build up the Inter Mountain Fair and its grounds. While I was there, the high school boys that needed a place to keep their project animals were allowed to keep them at the Fair Grounds.

In the late Fall of 1946 George Ingram who had just returned from the Service took over the Manager's job. He had been one of my top Agriculture students. I think I helped persuade him to apply, and then helped talk the Directors into hiring him. I am proud to say he helped Clair Hill develop the fine facilities that are there now and has promoted and conducted an excellent fair each year for the past thirty years.

Jesse Bequette moved on to another occupation, in another area, but returned to help celebrate the "FIFTY GOLDEN YEARS OF THE INTER-MOUNTAIN FAIR". Thank you Jesse. You will not be forgotten.

A BLUE RIBBON MANAGER

George Ingram

George Ingram started his career at the Inter-Mountain Fair by showing his project animals with the Lasta Paiute 4-H Club. His 4-H Leader was Maple Perkins and the Lasta Paiute club name came from Lassen and Shasta, as there were members from each county. That club evolved into the Cloverleaf 4-H Club of today.

Little did George expect to become the manager at that point in his life. George attended High School at Fall River and had Jesse Bequette as his FFA Advisor. After high school George went into the service. He was a secretary in the service as typing had been one of his best subjects in school.

After he was discharged he was in San Francisco visiting David Schneider who was in the hospital and a telegram was sent in care of David to George. "Come home, I have a job for you." Signed, Jesse Bequette. Although George was only twenty years old, he was offered the job of the manager for the Inter-Mountain Fair.

When George started his position with the fair in 1946 his only tools were a shovel and a hammer. He was offered $200.00 a month wages. This was big money, and how hard could this job be?

Frances Gassaway was still the volunteer Secretary, and Clarence Whipple was the Treasure. The Fair Directors were Willis Albaugh and Asa Doty. The fair ran smooth, no real problems. People were great to volunteer to help with the projects, and any time a job demanded more that George's shovel and hammer, he would just go over and borrow anything from a tractor to a wheelbarrow from the McArthurs.

In 1946 the Inter-Mountain Fair was changed to the Labor Day weekend and ran for three days. Jesse Bequette was the fair manager still as George did not take over until after the fair in the Fall.

The Committee Chairmen were, Livestock, Floyd Bidwell; Feed, Roy Hoffman; Domestic Science, Elizabeth Albaugh; Domestic Art, Hazel Kaufenberg; Flowers, Alice Burton; Police Protection, Roy Duggens and Clair Engle, Congressman.

The 1947 Fair was George's first fair and Frances was doing the job of both Secretary and Treasure by this time. Jim Bruce was taking care of the Livestock responsibilities and Della Wiertzba, Louise Crum and Anna McArthur were seeing to the Domestic fair exhibits.

George was lucky that he had the typing experience that he did, because to get the premium book approved a hand typed book had to be submitted to the state division of fairs and exposition to be approved before you could have it printed. George typed the entire premium book by himself and it was approved.

1947 was the first year for a carnival to be at the Fair.

1948 the fair needed to add to their real estate and the board approved the purchase of 97.776 acres from PG&E for the sum of $889.32.

In 1949 Hugh Carpenter was added to the board of directors along with Willis and Asa. The fair was running smooth and there were no real problems. This is the year some lots were bought from the McArthur family for $110.00. Also this year an exhibit building was started and a new grandstand was added to the fair grounds.

During the years of 1949 and 1950 the main hall, known as Ingram hall was constructed. For the grand opening of the hall a dance was held and all the ladies received an orchid corsage. The second race track and exhibit building were also ready for use.

The development of trees and shrubs started taking shape in 1950. The trees, shrubs and rose bushes were ordered from Stark Brothers nursery in Missouri and were sent by railroad to Alturas. Albert Kenyon was the first caretaker of the fair grounds, and he and George went to Alturas to pick up the order from the nursery. Albert and George planted all the plants and then planted the lawn. There was no sprinkler system, so George and Albert watered the lawns with the fire hose to get the grass started.

Some of the trees were donated and planted by members of the community as memorial contributions. Betty Eldridge and her 4-H group planted the spruce tree near the front gate that has grown into a beautiful large tree. The Glenburn Bridge Circle planted a tree in memory of Doris Brown. Willis Albaugh dug up some trees along the river and transplanted them on the grounds.

1952 in the directory for the Division of Fairs and Expositions the Inter-Mountain Fair was listed as the largest paying premium fair in California for beef cattle, and ranked the highest as the largest paying premium fair in the North State.

1954 the Junior Livestock exhibits were started.

About this time the flower gardens were instigated. To begin with the gardens were a ten foot by twenty-

George Ingram and Willis Albaugh, discussing their next job.

45

Albert Kenyon, the first maintenance man to help George plant all the lawn and trees.

Inspecting the new office building
in 1962, are
Sam Thurber, Willis Albaugh,
and George Ingram.

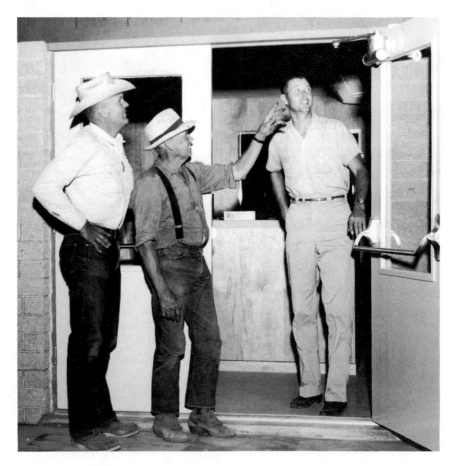

ABOVE:
Dennis Hoffman, Margie Urricelqui,
and Dan Marcum.

AT RIGHT:
Willis and George pouring cement
for the walk to the new office.

five foot garden. Later they were changed to a three foot by fifty foot flower boarder and were entered in to the competition by different individuals and clubs. This is also an unique way to beautify the fair grounds, and an idea that George discovered at another fair ground.

In 1958 the 40th Annual Fair came along and it was time to add to the Fair Board and Willard Brown and Wally Hilliard joined the group. Hugh Carpenter moved away so his vacancy had to be filled.

In 1962 the new office was built. A University of California Shasta County Farm Advisor had been established in the area and an office for him was built into the new fair office. Sam Thurber was the first farm advisor to occupy the office and he was there from 1960 to 1968, then Walt Spivy took over that position from 1968 to 1975. When he retired the powers at the University of California decided they could save some money by closing the McArthur office. The farmers and cattlemen in the area raised such Cain and did everything they could to keep this position filled. The University listened and in 1976 a young eager Dan Marcum took over the position of Farm Advisor in the area. Marcum is a plant pathologist with degrees from U.C. Davis and U.C. Riverside. In this area agriculture is big business. Dan is there to help the smallest gardener with problems with her petunias as well as the largest farmer with thousands of acres or hundreds of cattle. While Dan has been in the Inter-Mountain area a real growth in the agriculture crops has expanded. Such things as wild rice, garlic, strawberries, mint and more development in the hay and grain fields are attributed to his knowledge and leadership. Dan is an invaluable member of the community and is a tremendous asset to the fair. His boundless energy is overwhelming.

In 1965 Catherine Ryan replaced Asa Doty on the board. This is the year of the first queen pageant.

In 1967 the Junior Livestock Sale was started with Dick Nemanic as the President and coordinator. A position he still holds today.

In 1968 the Inter-Mountain Fair celebrated their 50 Golden Years of success. This is also the year that Gail Ashe started working as Secretary and Connie Addy retired.

1970 the covered arena was built.

1970. Covered arena being built.
L. to R. George Ingram and son Rob, Queen June Packham, Robert and Lois Ingram.

George Ingram (on the right) in recognition for 25 years of service to the fair industry awarded to him by Robert Finch (on the left) counselor to President Richard Nixon in 1971.

The Ingram family were all there to share happy moments with their dad, Fair Manager George Ingram.

George Brown left the fair board in 1972 and Leland "Bud" Knoch was appointed to the take his place.

1979 The old timers day was changed to the Golden Wedding Dinner, where everyone that has been married for 50 years is invited to share a wonderful dinner, entertainment and wedding cake.

1965 Catherine Ryan retired and Robert Thompson was appointed to the board from the Hat Creek area.

1974 Rose Schneider took over the job of keeping the grounds beautiful. This was her pride and joy and she did an exceptional job.

In 1983 George was President of the Western Fairs Association. He enjoyed this position and during that time traveled to some very interesting places, always being an impressive representative for the Inter-Mountain Fair.

Although it is true that Labor Day weekend is the time for the fairgrounds to shine and bustle, George, with the help of his staff, is kept busy thorough the year. It is not just a holiday time vocation. It has been his life. The fairgrounds is busy year round with weddings, dinners, parties, clinics, firemen carnival, health fair, 4-H events and horse shows and the Cattlemens yearly feeder sale or anything else that requires use of the buildings.

George was born and raised in McArthur and lives there with his wife. Their five children were raised on the ranch where George and Phyllis still live. The ranch was originally homesteaded by his great grandparents,

Gail Ashe and Dennis Hoffman.

the Straubs.

George retired from the manager position of the Inter-Mountain fair in 1988 after forty two years of dedicated service. He enjoyed every minute of his job and did it first rate. He was proud of his achievements as fair manager and the board was proud of his accomplishments.

So, to the man who gives a lot of Blue Ribbons, but never receives one himself, "One Big Blue Ribbon to George Ingram."

A HANDS ON FAIR MANAGER

Dennis Hoffman

On his fortieth birthday in 1988 Dennis Hoffman started a new career as Fair Manager of the Inter-Mountain Fair. A new beginning in a new area. Although he had passed through McArthur several times and thought it would be a neat place to live he never really thought it would happen.

At the time the job opening was announced when George Ingram was ready to retire, Dennis was the manager at the Madonna Ranch at San Luis Obispo, California. Raised in central Oregon at Redmond, Dennis had been in the purebred livestock business all of his life. It was a time in his life he was ready to make some changes and being fair manager was something he was quite familiar with. Dennis had a lot of friends in the fair manager positions, and they encouraged him to apply for the opening. The deadline for applications had approached and Dennis sent his in just in the nick of time. He came for an interview and the next day George Ingram called him and told him he was hired. It all happened so fast that Dennis was astounded, but very pleased.

Dennis is perfect for the Inter-Mountain fair as he wanted something that was a hands on job and not just a desk job. He mows all the lawns in the summer, because it gets him outside, away from the phone and gives him time to think. This is when he says he gets his best ideas. It also gives him the opportunity to know what needs to be fixed and either fix it himself or have his capable maintenance man take care of it. He likes the interaction with the community and the livestock people and knows how to talk their language.

Gail Ashe was still with the fair when Dennis started and he said he couldn't have done it without her. She was so very valuable in coaching him as to what needed to be done. Gail was there long enough to help Dennis learn the ropes then Valerie Lakey was hired in 1990 when Gail retired.

Jim Greer.

One of the things that Dennis is very proud of at the local fair is the fact that the Livestock Pen Show is the second largest Pen Show in America, second only to the Denver, Colorado Livestock Show. The facility is set up for the feeder sale pens and adapts well to the fair time activity. It also fits the communities needs as there are so many commercial cattlemen in the area. The halter classes are also well attended, but not near the magnitude of the pen show.

The remodeling of Ingram Hall was a major milestone for the fair recently. A building that is used for weddings, dinners and dances needing some major face lifting after forty-five years of heavy use. Everyone enjoys the new wall covering, lighting and stage.

The new R.V. Park is also a new addition to the fair which will bring in revenue for the fair. With fifty-three full hookups and rest room laundry and shower this facility is a real asset to the community. It was built with a revenue generating grant from the Division of Fairs and Expositions to help to stimulate revenue so fairs can become self supporting.

Dennis had some big footsteps to follow after the administration of George Ingram, but he is running right along full speed ahead, and takes his job very seriously. His ideas and willingness to be right in the middle of everything makes him a tremendous fair manager.

Jim Greer

The responsibility of keeping the grounds and buildings in good repair and maintaining them is thrown on the shoulders of Jim Greer. Jim whose has be handy man and maintenance man since 1985 has been given the new title of Structural Craftsworker. But still when something doesn't work or needs a little fixing it is Jim you call, no matter what his title might be.

The Women In All Their Glory

THOSE VALUABLE SECRETARIES

The Secretaries of the Inter-Mountain Fair have played an important part in the betterment of the fair. Frances Gassaway was the first secretary and was a volunteer from the beginning. A community minded citizen as well as a busy school teacher, Frances took it upon herself to take the minutes of the fair and later was also the treasure after Clarence Whipple turned that position over to her. During Fair time Frances was the only one to hand write all the entry forms and wrote out a tag for each item entered for exhibit. Frances received $50.00 a year for her position as secretary of the fair. A busy lady, but she enjoyed her involvement with the fair. Frances held the position of Fair Secretary for thirty years before retiring.

Connie Addy took over the position of fair secretary for a while and enjoyed that position.

In 1968 Gail Ashe took up the reins and responsibilities of Administrative Secretary of the fair and

1960. Clerks of the Inter-Mountain Fair . . . Top, L. to R., Joanne Beck, Hazel Bassett, Val Addy, Louise Clark, Carol Dustman, Freda Thorlakkson, Bottom, L. to R., Fair Manager George Ingram, Connie Addy, Grace Morris.

George Ingram, Fair Manager and Gail Ashe, Secretary.

George Ingram and Gail Ashe at the Western Fairs Convention.

worked under the management of George Ingram for twenty-three years.

Gail was not exactly the woman behind the scenes, because as anyone who had ever had any interaction with the fair office knew that she was a very visible part of what went on there. Although her state classification was "Secretary", the fair manager George Ingram regarded her as administrative secretary.

George and Gail worked more as a team to see that the fair ran smoothly. Gail was also the secretary to Dan Marcum, Farm Advisor for Shasta and Lassen counties. Dan said that Gail did everything and was a true asset to the community.

Gail has been involved with the fair most of her life. As a young girl she entered her cooking and sewing projects. While she was in high school she helped put the agriculture products on display. In 1967 when the fair needed someone to handle the horse show and swine exhibits Gail volunteered to do the bookkeeping and clerical work for both events. In 1968 Gail was hired to replace Linda Fava as the Secretary of the Inter-Mountain Fair and to work with George and Dan.

Gail's duties included handling the contracts for concessions, the year round rental of the grounds and renting the commercial space for the booths at fair time. Gail hired and fired the clerks and helpers during the fair. The queen pageant was also her responsibility which she very capably handled.

After George Ingram retired Gail was there to help the new fair manager Dennis Hoffman learn the ropes. Dennis said it would have been really difficult for him if Gail had not been there to lead the way. With a smile and a gentle shove she showed Dennis the things that needed to be done without ever trying to be the boss.

Gail retired in 1991 to do some of the things she missed out on during those busy years that she so competently handled the affairs of the fair office.

Good job Gail, you were one in a million.

Valerie Lakey took over the responsibilities of Secretary in 1990. Val is actually classified as the Business Assistant. Her duties are so numerous that it would be difficult to write a job description for her. She is so efficient in her position that it would be hard for anyone to fill her shoes.

At least the fair secretary does not have the responsibility of the Farm Ad-

visors office as in the past. Margie Urricelqui very capably handles that job.

Kris Cunningham works as the Exhibit Supervisor during fair time. She handled the horse show clerking for twenty years. Bonnie Sattler is another strong supporter of the fair and has worked there for several years at fair time.

The fair now hires fifteen clerks, fifteen gate people, five night watchmen and five extra maintenance people during fair time.

ROSES FOR ROSE

Did you ever see a rose any prettier than the ones that grow at the Inter-Mountain Fair Grounds. Morning Glory, corn stalks and sun flowers hiding the old metal siding of an exhibit building, making it a thing of beauty. Petunias planted around a fire hydrant or marigolds at the base of a light pole. Who would ever thought of planting sweet peas here or zinnias there. Only Rose Schneider would. Not a bare spot is left for an ugly weed to grow, because Rose doesn't give a weed a chance. She plants a flower seed or plant there before a weed has even has a chance to wake up in the spring and take root. Rose surely must lay awake at night planning her garden beds at the fair ground, because when spring calls, Rose was there to start digging in the dirt.

Rose Schneider started doing the flowers and lawns at the fairground in 1974. Born and raised in McArthur,

Tammy Kofford, Photographer and Valerie Lakey, Business Assistant.

Rose Schneider tending her gardens at the fair.

the fair has been a part of her life every year. Her pride in her home, town and fair is unmistakable. From daylight till dark from May to the heavy killing frost, Rose was on duty to plant the seeds and plants, change sprinklers, battle bugs and destroy weeds. An ambitious lady with a love for beauty in flowers and plants, she shared her passion with others so they might enjoy the most attractive fair grounds in Northern California.

The flower borders that line the walk way of the fair grounds are part of the fair competition that is entered yearly by organizations and individuals of the community.

Rose retired from her duties in 1993 but thank you Rose for a job well done. Your beauty in flowers will last forever.

THE ROYALTY OF
THE INTER-MOUNTAIN FAIR

The Queen contest of the fair was started in 1965. This pageant was open to any girl from Burney, Fall

Gail Ashe, Rose Schneider, Bonnie Sattler.

River or Big Valley High Schools. The girls must be from 16 to 21 years old and have never been married.

The Queen contestants are sponsored by a business, organization or individual. The sponsors contribute to the expense the queen must incur to provide them for the proper clothing that is needed for the event.

As with most of the events of the Inter-Mountain Fair a lot of volunteers help with the queen pageant.

On a summer evening in July the pageant is held at the fair grounds and followed by a dance. The contestants are judged on beauty, poise, personality and communication.

The responsibilities of the chosen queen is to promote the fair. The Queen and her court visit other fairs in the area and ride in the parades to publicize the Inter-Mountain Fair. During fair time they participate in the various events by passing out ribbons and lending their support where needed.

The Inter-Mountain Fair Queen contest was started in 1965 and the queens were as follows.

Kristine Holl	Big Valley	1965
Janet Oiler	Fall River Valley	1966
Leslie Thompson	Fall River Valley	1967
Debbie Ewin	Fall River Valley	1968
Sharon Telford	Fall River Valley	1969
June Packham	Fall River Valley	1970
Mary Harper	Big Valley	1971
Patricia McArthur	Fall River Valley	1972
Lani Bickle	Fall River Valley	1973
Norma Oiler	Fall River Valley	1974
Stephanie Wade	Burney	1975
Vicky McArthur	Fall River Valley	1976
Kim Hines	Fall River Valley	1977
Sherry Dean	Fall River Valley	1978
Terri Drewry	Fall River Valley	1979
Cindy Campbell	Fall River Valley	1980
Donna O'Connor	Fall River Valley	1981
Lori Dean	Fall River Valley	1982
Noelle Barrington	Fall River Valley	1983
Lori Brown	Fall River Valley	1984
Kristy Klagues	Fall River Valley	1985
Heidi Covington	Big Valley	1986
Stephanie Rowe	Fall River Valley	1987
Trina Tuschen	Burney	1988
Cory Thomas	Big Valley	1989
Jenny Britten	Big Valley	1990
Julie Copp	Big Valley	1991
Marci George	Big Valley	1992
Megan Ray	Fall River Valley	1993
Stacia Hunt	Big Valley	1994
Heidi Buckman	Fall River Valley	1995

1966 fair queen Janet Oiler on left.
1965 fair queen Kris Holl on right.

1967 fair queen Leslie Thompson.

1970 fair queen June Packham.

1969 fair queen Sharon Telford.

1971 fair queen Mary Harper.

L. to R.: Reba McCord, 1972 fair queen Patricia McArthur, Shirley Kennemore.

L. to R.: Debbie Radle, 1973 fair queen Lani Bickle, Debbie Vigelo.

1974 fair queen Norma Oiler.

Left: Pam Thompson, center: 1975 fair queen Stephanie Wade.

L. to R.: Rena Oilar, 1976 fair queen Vicki McArthur, Sherry Grames.

1977 fair queen Kim Hines.

1979 fair queen Terry Drewry.

1984 fair queen Lori Brown.

L. to R.: Rachelle Knoch, 1987 queen Stephanie Rowe, Susie Thompson.

1989 fair queen Cory Thomas.

L. to R.: Stacie Nordman, 1991 fair queen Julie Copp,
Jody Chandler.

1990 fair queen Jenny Britten.

L. to R.: Kimberly Bell, 1992 fair queen Marci George,
Sheila Walker and M.C. Cal Hunter.

ABOVE LEFT:
1993 fair queen Megan Ray.

ABOVE RIGHT:
1994 fair queen Stacia Hunt.

AT LEFT:
1995 fair queen Heidi Buckman.

Through the Years

In the lives of every child, clowns, ferris wheels and fairs are a wonderful thing. In the lives of every grown up in the Inter-Mountain area the fair is also a wonderful experience. Whether you participate by entering your favorite fudge or cake, a flower, quilt, animal or picture, it can be a rewarding experience.

If you are not one to enter, then surely you can be the one that enjoys the craftsmanship of others and oh and ah at the beauty of your friends' labors.

There are so many things to enjoy at the fair. Rodeos, horse shows, logging shows, destruction derby, concerts, fashion shows, pet parades, sheep, hogs, cattle and horses. The biggest pumpkin or squash, the craziest scarecrow, the most beautiful art work and quilts, cookies, cakes or that very best jam. Needlework, woodwork, flowers and vegetables galore. The 4-H booths, commercial booths, cotton candy and corn dogs, it's all there at the Inter-Mountain Fair of Shasta County.

There is something for everyone even if it is setting in the shade sipping lemonade and watching people stroll by.

The hustle and bustle of all the excitement, the carnival music, that magic ride on a ferris wheel, watching the excitement of a child on a merry go round.

This is what the community had in mind in 1918 when they started that first rodeo and exhibits. But never in their wildest dreams could they imagine the tremendous growth the Inter-Mountain Fair has made and the fantastic entertainment it brings to the community every Labor Day weekend. For seventy-seven years the fair has enjoyed continuous growth and just gets bigger and better every year.

As told in the beginning a handful of people got together and decided to put on a fair. Each year improvement was made and a series of newspaper articles tell of the advancement.

Sept. 26, 1919, "BIG CROWD COMING TO OUR FAIR, OCT. 2-3-4". This is the big event, the first Farm Bureau Fair to be held in this section of the country. Ladies of the Hat Creek Farm Bureau suggested that each and every member take or send something to be displayed at the fair. We all must contribute if we expect to have a fair and everyone was urged to attend the event.

In 1920 it was noted that Scott McArthur donated a tract of land near McArthur to be used as a permanent location for the Fall River Valley Fair. The committee in charge of the fair will now proceed with plans for fences, buildings, and other improvement, which will be needed to prepare the tract for use.

The task then began to raise money for the erection of grand-

Emile Estes on Merry-Go-Round ride.

ABOVE:
A 4-H booth.

AT LEFT:
That beautiful fancy work.

BELOW:
The Wilcox Family Farm booth.

The Art exhibits.

stands. This was accomplished by selling ten year admission tickets and part of the grandstands were built.

In 1926 the exhibit building was built and exhibits were moved from the Rose barn. Grange members helped build the building with lumber from the old grandstand in return for use of the building for their meetings.

Forrest Loosley served as president during 1925 with Willis Albaugh as secretary. Officers in 1926 were Marple Laird, school principal and Willis Albaugh again serving as secretary. Money was raised during these years for the fair through dances and other activities in the area and Big Valley.

In 1927 the McArthur Grange assumed the management portion of the fair due to disinterest in handling of the affairs. They, through officers and committees, took an active part until the early 1930s. George Brown, Leland Anders, Les Agee, Frances Gassaway and Byron Hollenbeak served as association officers. The help of the local high school agriculture teacher was enlisted to set up and care for the agricultural exhibits.

Secretary Willis Albaugh said that in the early years the funds of the fair was kept in a cigar box and the cowboys at the rodeo were paid for their rides directly out of the box.

Large crowds viewed the exhibits at the 19th annual fair in 1936. Although the weather was very unsettled, the crowd enjoyed the many diversified products of this area on display. Exceptionally good this year were the displays of local school students, which comprised many forms of workmanship including paintings, crayon work, sketches, composition and many other good specimens of school work.

An interesting exhibit of handiwork including engraving, bird mounting, carving, whittling and woven belts was the contribution of the local CCC boys to the show and this attracted considerable favorable comment.

Chester Bethel's big pumpkins were on display as usual, and many other fine displays were on exhibit from local gardens and fields.

The boys of the FFA under the supervision of Jess Bequette played a prominent part in the agricultural and livestock show. Walter Boles was in charge of the booth

65

1937. Cale Guthrie ready for the Grand Entry.

1937. A clown with his trick bull.

1947 horse show.
The old grandstand was converted to the hog barn and the arena was where the office now sits.

1947. Inter-Mountain Fair Rodeo Grand Entry.

for this organization. Elmo Brown exhibited a fine display of vegetables, Harry Brown won sweepstakes with his exhibit of hogs, Robert Bruce cleaned up the blue ribbons with his sheep and Jim Selvester took high honors in the grasses and forage crops exhibit.

In the poultry division Nellie Knoch, Walter Callison and David Shaffer were winners. In Tom Turkeys, David Shaffer and Robert Bruce placed. Hen Turkeys were won by Walter Callison and David Shaffer. The best white eggs were displayed by Nellie Knoch with Lora Collett second. The brown egg division was won by Frances Hollenbeak and Walter Boles.

Robert Bruce had the first place ewe in the Sheep division with Nellie Knoch as second. Leland Anders had the best Buck with Robert Bruce and Harold Burr running right behind him. Bethel Brown had the Champion Boar in the swine division, Percy Creighton, Roy Bassett and Ben Gassaway also placed in this department. The honors of having the best sow went to M. Baum, Elmo Brown, Ralph Opdyke and the Gilt champion was won by Harry Brown, Mary Baum and Roy Bassett.

The Dairy cattle division was won by M. Baum, Arthur Phillips, Leo McCoy and Fritz Edmonds.

Beef cattle bull was won by Jim Day with Ben Gassaway having the best cow. The heifer honors went to Della Wiertzba and Bill Wiertzba.

The horse show division was well attended with the Draft Stallion division won by Harold Burr, Percy Creighton and Hardy Vestal. The Saddle Stallion honors went to Byron Hollenbeak, and Jess Eldridge. Draft Mare first place belonged to Robert Shaffer and second Leland Anders. Draft team honors went to Jim Day. In the colt division, Mary Baum, Cale Guthrie, Leland Anders, and Lily Bognuda were named as winners.

In an action crammed afternoon marked by hard riding, thrills and spectacular spills, Marshall Flowers of Anderson last Sunday took first money at bronc riding in the annual rodeo of the Inter-Mountain Fair. In spite of cold winds and generally threatening weather conditions, a good crowd practically filled the grandstand to witness the bucking show.

In the bucking contest, Sie Elliott and Eddie Hess followed Flowers, to take second and third money respectively. In the free for all horse race, Nelson Monroe took first money with Ted Bazin riding second. Ted Bazin won first prize in the saddle horse race, and Arthur Cessna was tops in the mule race. Lily Bognuda, the Little Valley cowgirl won the stake race and Jimmie George took second place. Relay honors were won by Lily Bognuda, with Jimmy George in second.

The steer roping event was won by Ralph and Tom Evans, with Jess Eldridge and Henry How in second.

The latter event was featured by a spectacular and near tragic accident when the galloping horse of Buck Richardson of Tehama County stumbled and pitched over end and rolled over its rider. When the dust cleared Richardson was lying prone and lifeless on the ground. A small crowd quickly gathered and artificial resuscitation was begun and the unconscious man was finally brought to after several minutes. Spectators pronounced this one of the best shows ever to be given in the McArthur arena.

In 1937, Buster Ivory took his first ride in a rodeo at McArthur, not far from his birthplace at Alturas. He won first place in the cow riding and steer stopping and pocketed $58, roughly 58 times his daily wage as a ranch hand. Ivory was only fourteen years old and that day his career began. Buster Ivory hit the rodeo circuit and rode bulls and broncs. After a severe accident, Ivory came back to be one of the five top bronc riders in 1951 and 1952. For the next 30 years Ivory continued in the rodeo business as a contestant, judge, livestock superintendent, arena director, chute boss, manager or producer.

In 1978 Ivory was "Rodeo Fan's Man of the Year" and was inducted into the Pro Rodeo Hall of Fame in 1991. It's quite an honor for an old country boy who got his start when he was fourteen at the McArthur Rodeo.

1940 papers tell of the new grandstand being constructed for the fair. A crew of carpenters are working feverishly to complete the addition of a new section of the grandstand. When finished the seating capacity will be doubled so that every spectator can be assured of getting a ringside seat.

Considerable work is also being done on the stock pens, shed, corrals and pens. By the opening day, Saturday, September seventh, all of the improvements will have been made, according to Director Jesse Bequette.

At every rodeo and fair in every Northern California city and town, the McArthur wild west show has been widely publicized. According to the response already shown the attendance at the Rodeo this year will exceed the numbers attending in the past decade. Many of the same performers who are seeking action this year at the Lassen County rodeo, the Redding Round-up and the Siskiyou county fair, have been signed to ride here. These buckaroos, who follow the various shows throughout the country will put plenty of professional punch into the show this season.

Lay away your bib and tucker! Don your spurs,

chaps and high heeled boots and prepare to kick over the traces, for we're all McArthur bound. Today marks the beginning of Eastern Shasta county's biggest attraction, the Twenty-Third Annual Fair and Rodeo.

Beyond any question of a doubt the rodeo, which is to be held on Sunday afternoon, will be one of the best wild-west performances ever held in the district, all of which is assured by the predominance of two vital factors that go to make a good show, good spirited mounts and good riders.

The best riders will be here. Those who attend the show Sunday will not only see good riders, they will have one big treat in sighting the best riders of the nation in action. Here are just a few of the riders that will compete for honors; Perry Ivory, Bob Locey, Jack Myers, Dutch Bartrom, Sie Elliott, Leonard Johnson, John Schneider, John Bowman, Cuff Burel, Chuck Sheppard, Gene Rambo and many others. It was indeed gratifying to the Rodeo committee, Bill and John Lee and Roy Cessna when these boys sent in their registration money.

Big money is being offered this year. In the bronc riding event alone, $70, $35 and $15 is offered for first, second and third respectively is being offered with the entrance fee added. This will put the final purse up to worthwhile money.

Equally good money is being given in the calf roping, cow riding, bulldogging and other fast events.

Among performers will be the clever roping team from Lassen County. This trio, consisting of Pierce McClelland, Clint Haley, and Tommy Johnson, promise to bring applause from the audience when they step into the arena as a unit to do miracles with their lariats. It will be worth the price of admission to see a quintette of champions in action.

For several weeks the committee has been rounding up the best bucking and race stock in the northern country. Sixty head of the best animals will be eager for the "take off". Jack Metzgar's string of race horses as well as the McGonnigal horses from Mt. Shasta will be pitted with other fast animals of the district. As a climax to the show a special pony express race is to be featured, teaming the best animals at the fair this year.

McArthur to have Fair and Rodeo September 6 and 7, 1941.

Premium money of $8,729 is being offered to exhibitors. Premiums have been greatly increased in nearly every department with many new departments added. There was an increase of from $1,500 to $8,729 over the 1940 fair.

One of the new attractions offered is the range cattle classes to be penned in the new corrals built for

1947 Horse Show. Bill Carpenter driver, Hugh Carpenter passenger, Fred Bayliss judge.

69

Jim Dimmick, 1950.

that purpose. Premiums allotted in that division total $482 for steers and $228 for heifers.

A new 32x96 stock exhibit barn is under construction and a water pressure system are among the 1941 improvements.

The premiums offered are divided as follows: Draft and light horses, $808; mules, $96; Beef Cattle, $1530; Dairy Cattle, $1368; Swine, $666; Sheep, $332; Poultry, $99; Honey, $11; Grain, $106.50; Grasses, $96; Vegetables, $164.50; Fruit and Berries, $40.50; Cut Flowers, $98; Canned and Baked Foods, $125; Needle work, $159.25; Art, $96; FFA, $1977; 4-H Club, $385; Agriculture booths by farm organizations $225; and Horse Show, $345.

Agriculture Booth entries accepted for displays by farm organizations whose principal and primary purpose is for the furtherance and improvement of Agricultural production. Premiums for such booths are as follows: 1st, $75; 2nd, $60; 3rd, $50; 4th, $40.

Anyone wishing information and premium books may contact Secretary F.N. Gassaway at Pittville, California.

After the fair on Saturday night the Inter-Mountain Fair Association is sponsoring two dances to accommodate the crowd. The McArthur Grange Hall will feature Ted Corder and his Lassen Cowboys with cowboy music. At Fall River, the Town Hall will feature Jimmie McDonald and his orchestra, featuring modern music.

Sunday afternoon the Rodeo will top off the fair and close the weekend. Arrangements for the rodeo will be in charge of Bill Lee Jr. and as yet are not complete.

The 1941 records show that Jack Meyers won 1st in Bronc Riding for $122.50, Leonard Johnson won 2nd for $66.50 and Loren Carpenter won 3rd for $36.00. Jess Eldridge won 1st in steer stopping for $26.25 and Jim Eldridge and Cris Emeric split 2nd and 3rd for $11.62 each. Albert Albaugh won first in the Mule Race with Bob McGongale taking second. A man was paid $1.00 for renting a cow to the fair for the activities or $2.50 for the loan of his horse.

The 1944 fair held on October 21 and 22 was very successful. The largest livestock exhibit ever shown at

71

Forage and
Vegetable
displays.

that fair. The range cattle division was outstanding, with Perry Opdyke's fine herd of Hereford cattle winning first in the pen of ten steers, and first and second in the pens of five heifers. Mr. A.E. Royce, former owner of the Fall River Meat company won first with a pen of five Angus heifer calves. Mr. Royce also exhibited the grand champion fat steer of the show which was one he had purchased during the summer from Goldie Wilcox of Hat Creek. Floyd Bidwell, Cassel, exhibited the grand champion bull and cow of the purebred Hereford division. The other honors were about equally divided between John McArthur, Floyd Bidwell and Wixon and Crowe Ranch of Millville.

In the Future Farmer range cattle division, Gene Bidwell won first in the pen of three steers with the McArthur chapter's own steers. The Spalding Brothers won second and third. Ernest Spalding won first in the individual class, with Gene Bidwell's steer placing second and third. Charlie Cessna won first and second with a nice pair of Angus heifers.

Kenneth McArthur exhibited a fine group of beef Shorthorns to win all the prizes in this division. Although he had no competition this herd is an excellent group, having originated from excellent bred Canadian stock. Morris Doty won nearly all the prizes in the milking Shorthorn division, with Lawrence Day running him a close second. Ed Albaugh won many blue ribbons with his fine group of Berkshire hogs.

George Strickland's team won the horse pulling contest, with John McArthur placing second and George Brown third. In the draft horse division John and Kenneth McArthur won most of the blue ribbons.

In the Future Farmer beef showmanship contest, Kenneth McArthur won first with his beautiful little roan heifer that Judge Bayliss of Hilt, California, pronounced as being an exceptionally fine animal. John Meeker won second with his little blue-ribbon Future Farmer Hereford bull. Charlie Cessna placed third and Gene Bidwell fourth.

On Sunday afternoon the Young Farmers Association of the Fall River High School put on a sensational rodeo. In the roping event first went to Jack Conlan, second to Lil Bognuda, second to Jess Eldridge and third to Hank Stone. Saddle horse quarter mile race, first to Floyd Bidwell and second to Arthur Cessna, Pony race, first to Eddy Meeker and second to Sammy Eastman.

Purchase orders dated 10-24-44 and approved by Jesse Bequette were as follows; J.W. Bequette, 1 year fair manager, $100.00; Frances Gassaway, 1 year fair secretary, $50.00; John McArthur, five and one half days labor repairing fair ground fences, $27.50; Ken-

neth McArthur, two days use of tractor and wagon to clean up fairgrounds, $10.00; McArthur FFA Chapter, six and one half days cleaning up fair ground after fair and preparing grounds for winter, $26.00.

2,500 ATTEND McARTHUR FAIR IN 1947. With a crowd of 2,500 people in attendance, the 29th annual Inter-Mountain Fair proved to be a success. A large entry list competed for the more than $25,000 put up in premium money, with the livestock section being especially well represented.

Sunday's rodeo, put on by the McArthur Grange under the direction of Bill Lee, saw the grandstands, which hold 1,200, overflow and hundreds stood around the arena to watch the thrill packed program. A good sized crowd turned out for yesterdays horse show, while dances Saturday and Sunday evenings were well attended.

One of the big features of the horse show was the victory scored by Bill and Bob Carpenter's horses in the heavy pulling contest. Their horses were the only ones to pull 3,600 pounds the required twenty feet.

The horse race was mighty exciting, especially for Hardy Vestal. The horse he had in the race won first, and Hardy got so excited that he jumped off of the grandstand and broke his arm.

Managed by youthful George Ingram, in his first season in that capacity, the fair offered many well arranged exhibits. The McArthur Grange, showing a large variety of products of that area, was the first prize winner in the feature display division, while the Hat Creek community was second and McArthur Farm Bureau third.

In the junior feature display division, the Soldier Mountain 4-H boys and girls of Glenburn both won first prizes. The Fall River Mills 4-H girls took second.

Sunday's rodeo was started off with a big parade of all contestants including a covered wagon hauled by a tractor, the local fire trucks, a girl scout troop and Jimmy Dixon of Alturas, rodeo clown riding Fibber McGee, his trained mule.

A carnival offered amusement to the crowd in between events.

Willis Albaugh, president of the board of directors of the fair was in general charge of the horse show. Other board members are Asa Doty, vice-president; Frances A. Gassaway, secretary-treasurer and George Ingram, manager.

Jimmie Dixon provided some clever clown entertainment for the rodeo filling in lulls in the program with his antics, while Don Ray's orchestra, which played for both dances provided music at odd times. Al Meeker and Fred Ball announced the rodeo, while

Meeker handled the horse show announcing alone, filling in dull spots with a running file or comment on stock and performers.

Several rodeo riders were injured, marring the enjoyment of the rodeo. Hack Lambert suffered possible rib fractures when thrown in the bareback riding and was taken to a hospital. Bob Williams suffered a leg injury, while all three Carpenter brothers, Loren, Forest and Spike suffered injuries in riding events. Donald Matson's back was injured.

Rodeo judges were Sie Elliott and Marshall Flowers, while Joyce Turner and Bud Kaufenburg were timers. Fred Bayless of Hilt was horse show judge, as well as light horse division judge. Other judges were; feature exhibits, agriculture, horticulture, C.A. Kettlewell; livestock, George Bath; domestic arts and science, Mrs. Barbara Simpson.

Many familiar names were in the line up of winners in the Agriculture and Horticulture Department. Carl Hillman, Theresa Kolb, Mrs. C.M. Bidwell, Marian Totten, Albert Albaugh, Ivadell Carpenter, Kathryn Ingram, Walter Callison, George Brown Sr., C.C. Hollenbeak, Rose Agee, Ed Bickford, Evelyn Eldridge, H.B. Turner, Wanda Callison, H.P. Struble and Ollie Neuerburg. Many of these people had entries in every division, doing their best to make a good fair and lots of competition.

In the Domestic Science division awards were given to Mrs. C.C. Hollenbeak, Hazel Bassett, Mrs. C.M. Bidwell, Maxine Summers, June Boster, Mary Lee Carr, Elizabeth Albaugh, Norma Callison, Mrs. A.L. Doty, George F. Brown Sr., Shirley Neuerburg, Ollie Neuerburg, Rose Agee, Catherine Hillman, Mrs. J.H. Ryan, Ramona Cates, Jeanette Murray and E. Hall. Many winners were duplicated in many of the divisions.

The Livestock Awards listed for the 1947 fair were as follows; BEEF CATTLE. Aberdeen Angus; Charles Ryan, Anderson, four firsts, eight seconds, two thirds. C.M. McDowell, Orland, nine first five seconds, one third. HEREFORD; Floyd Bidwell, Cassel, 14 firsts, two seconds, Lem Earnest, Cassel, one second. SHORTHORNS; John McArthur, McArthur, three firsts, one second; Jean Albaugh, Adin, two firsts, one second; Dale Albaugh, Adin, one first.

HEREFORD LOCAL. Lem Earnest, Cassel, thirteen firsts, three seconds; John McArthur, one first, four seconds, three thirds. ABERDEEN ANGUS, LOCAL; Morris Doty, Cassel, fourteen firsts, two seconds . . .; Robert Conner, Anderson eleven seconds. SHORTHORN, LOCAL. Kenneth McArthur, McArthur twelve firsts; R. and A. Conner, Anderson,

one first; Charles T. Ryan, two seconds. HEREFORD FEEDER. Floyd Bidwell, R.E. Bartell, Al and Jim Bruce, W.R. Carpenter. SHORTHORN FEEDERS John McArthur. FAT ANIMALS. Floyd Bidwell, Morris Doty and Charles Ryan had the Champion Fat Animal.

DAIRY CATTLE. Winners, Holstein; Lem Earnest; Milking Shorthorn; E.L. Bixier, Morris Doty; GRADE ANIMALS. Charles Ryan, Floyd Bidwell, Jerry Dimick. Swine; Norman Hawkins. Sheep; Lester Agee.

JUNIOR DIVISION FFA. Beef Cattle, Billy Albaugh, Showmanship. HEREFORDS; Mark Totten, Wayne Eldridge, Bobby Lee. CHICKENS; Sara Horr, Gary Highley, Rose Agee. SHOWMANSHIP DIVISION: Beef Cattle; Keith Stubblefield, Dale Albaugh, Joann Bidwell, Betty Wilcox, Frank Giessner. Dairy Cattle; Jimmy Albaugh, Shirley Neuerburg, Johnny Hoffman, Phyllis Totten, Jackie Horr, Grace Brown. Swine; Malcolm Graeber. Sheep; Lawrence Agee, Rose Agee. Light Horse; John Silva, Bill Lee, Ulch and Stewart, E.B. Coffin, Jesse Eldridge, Harold Bidwell, Lil Lambert, Fred Ball and Harold Vestal.

In 1947 the Golden West Carnival show will be an added event to the Inter-Mountain Fair. With five big rides and various concessions and entertainment, it is promised to be a splendid addition to the excitement and fun. Owned and managed by Harry Fisher, he promises a thrill for young and old alike with numerous spectacular events.

Several buildings were added to the fair ground in 1947. The Quonset buildings were purchased from a construction site and moved to the fairgrounds. Several other buildings were acquired from the same project and either sold to reimburse the fair or used for other purposes. Before this time the fair grounds only consisted of the Jr. Exhibit and office building, the old grandstand had been converted into the livestock barn, three small barns and a set of bleachers.

In 1948 plans are in the making for a $70,000 exhibit hall for the Inter-Mountain fair at McArthur. George Ingram told the county supervisors that he hoped the work on the hall could start this year. Meantime the fair group is purchasing a 97 acre site from P.G.&E. for $825. Ingram presented the budget to the board which accepted it. According to the budget, total capital expenditures are expected to amount to $76,825. This includes the proposed exhibit hall, land site $1,500 for an eighty foot water well, $1,500 to level the land $2,000 for a half ton pickup truck and $1,000 for a wagon and miscellaneous articles.

There were seventeen pens of feeder cattle with 85 head in the range cattle show. 4-H and FFA entries total 35 beef and dairy cattle in the Junior Division. The senior department booths include Fall River Grange, Hat Creek Farm Bureau, McArthur Farm Center and the McArthur Grange.

A total of $9,369 was awarded to the 170 exhibitors at the fair with 1500 entries in all departments.

Approximately 2000 people viewed the exhibits and feature events.

The rodeo was wild and woolie and one of the main features of the fair. Bill Lee, rodeo manager and Don Gomez arena director kept things moving and the rodeo was the first to end before dark in the last several years. Pick up men Art Cessna and Orville Cessna were kept busy.

This year was a local rodeo with local riders. Although the horses were some of the wildest ever seen locally, the ambulance had to be called once when Jim Nunes was thrown by a saddle bronc.

About sixty-five horses appeared in the parade of which Jess Eldridge did an excellent job as parade director.

Western music was furnished throughout the afternoon by Don Ray's orchestra, with solos by young Carla Spalding of Dixie Valley.

Some of the outstanding rodeo events were as follows: Saddle bronc; Bob Conner, Phil Stevenson. Calf roping; Chub Coffin, Hugh Carpenter, Bob Jones and Jim Short. Team roping; Bob Jones and Jim Short, Chub Coffin and Bob Rosser, Jim Short and Bob Woolery, Chub Coffin and Dorothy Coffin. Bareback riding; Irving Wood, Bob McLoughlin, Bob Conner and Gaylord Thistle. Saddle cow riding; Jim Short and Irving Wood, Kenneth McArthur and Charlie Cessna, Jimmie Snipes and Arnold Jutten.

The community and organization booth winners were McArthur Grange with a theme, "The World Looks to the Farmer". The Hat Creek Farm Bureau's booth of "Peace and Plenty" took second. Third place went to the McArthur Farm Center which featured "California Gold", a painting of a miner in the background with all locally produced fruits, vegetables and grains in the foreground.

The Hat Creek 4-H won the blue ribbon with their booth on the theme of "Control of the Cattle Grub".

The main fair hall known as George Ingram Hall built in 1948-49.

The Fall River 4-H was awarded the red rosette for their display of handwork of the members, clothing, canned foods and baked goods, all produced by the members. The Soldier Mountain club was represented by two booths, the girls had a display of the progress of home food preservation while the boys theme was "Farm Safety". McArthur FFA and Modoc FFA also had booths. The one family farm feature exhibit of the fair was entered by the Herb Totten family and they won a first place.

A special prize this year is the trophy which shall be awarded to the best crocheted article of the fair. This award is being given by the National Needlecraft Bureau of New York which is sponsoring the nationwide crochet contest. Winning exhibits in the Inter-Mountain Fair's crochet contest will be eligible for the entry.

According to the judging sheet the following were the riders in the Children's Saddle Mounts. Billy Lee on Scotty, Norman Lee on Butch, Joe Fagan on Elmer, Gail Carpenter on Theaba, Chester Ball on Snoopy, Ron Elliott on Pepper, Punky Rice on Tillie, Bill Cessna on Beauty and Billie Koffard on Brownie.

Light Weight Trail Horses entered were Lee Hammon on Champ, Jim Bidwell on Big Valley, Frank Swords on Bisket, Bill Lee on Midnite, Harold Vestal on Silver.

Stock Horses entered were Claudia Lee on Jubilee, Roy Cessna on Smokey, Bing Rice on Lucky, Jess Eldridge on Fox, Spike Carpenter on Betty.

Children Saddle Mounts 14.2 and under Barbara Carpenter on Rusty, Warren Bell on Stormy, Lorraine Maxim on Shasta Oro, Don Elliott on Buttons, Jim Bidwell on Flicka, Leanna Spalding on Black R-O and Ronnie Kofford on Cheta.

At the September 1949 fair approximately 4,500 people attended. About $15,000 was paid out in premium money. There were 274 exhibitors at the fair and many new records were set. The exhibits were much more effective as there was plenty of room for all. The $165,000 worth of new construction that has gone into the fair building resulted in a much better fair. This money was supplied by the state through pari-mutuel betting at horse races and the only local tax money was $500 used for advertising. The new main hall is a welcome addition and will be enjoyed by the community for various events.

The 1950 fair has a new event to add to the entertainment of fair spectators. A log bucking contest will be featured Sunday morning before the grand entry of the rodeo takes place. The rodeo this year is sponsored by the McArthur Grange. The Horse Show will fill the Monday afternoon slot. Two dances will be held Saturday evening while one more dance will be held Sunday evening.

Acel Oilar won all the money in the Corriedale class at the fair in 1950 with his purebred Corriedale sheep. Kenny Eastman and Wayne Carpenter entered their range sheep in the FFA range class and also won. This was the biggest turnout that the FFA section has had at the local fair for a number of years here.

1952 was a banner year for the Shasta County Fair at McArthur as listed in the Report on Operations put out by the California Department of Finance. This year the McArthur Fair was the highest paying fair in the state in the Beef Cattle division. Sixteen exhibitors had 163 entries with 274 animals and the total awards paid were $5,435.00. The second highest fair was the Plumas Fair at Quincy with $1,300.00 less premium money than McArthur with more exhibitors but less animals.

That same year the Home Economics department had 92 exhibitors with 1048 entries and $1,433.00 in awards. The floriculture had 44 exhibitors with 287 entries and $581.00 in premiums paid. The total of all awards paid at the Inter-Mountain Fair in 1952 was $14,436.12.

The roster for the 1954 horse show, Heavy Weight Stock Horse, 1st Nelmar Spalding on Topaz, 2nd Hardy Vestal on Buck, Marion Carmichael on Bud. Children's Trail Horse, (listed in order of ribbons) Bobby Milner on Rusty, Wendall Carpenter on Micky, Tommy Vestal on Toby, Carolyn Tyler on King, Elaine Wiertzba on Cindy, Leanna Spalding on Wanda V, Warren Ball on Dusty, Charles Kramer on Paint, Sharon Carpenter on Judy, Billie Cessna on Beauty, Carla Spalding on Crescent and Bobby Milner on Rusty.

Light Weight Stock Horse Speed and Handiness Class (listed in order of placing) Hardy Vestal on Buck, Nelmar Spalding on Topaz, Nelmar Spalding on Roanita, Jesse Eldridge on Fox, Nelmar Spalding on Smokey, Hack Lambert on Fritz, Lil Lambert on Chip, Hack Lambert on Betty Lou, Charles Fagan on Jim.

Trail Horse (Open as to Weight) listed as to placing, Karl Hemstead on Kit, Johnny Milner on Flicka, Joe Fagan on Elmer, Lee Hammon on Champ, Arthur Cessna on Chub, Nelmar Spalding on Cap, Elbert Lee on Joker, Warren Ball on Prince, Wayne Eldridge on Snooper, Roy Cessna on Jess, Marion Carmichael on Bud, Ron Kofford on Jeff McGue.

Family Groups, 1st, Nelmar Spalding, 2nd Claudia and Elbert Lee, 3rd Bill Lee.

Pulling Contest light weight, John McArthur 1st,

Hugh Carpenter 2nd. Heavy Weight, 1st Hugh Carpenter, 2nd John McArthur, 3rd Karl Hemstead.

All the grounds around the new exhibit building have been planted to alfalfa and grass. This will add to the comfort and pleasure of the thousands of people expected to attend this year's three day fair. The roads have been graded and rolled. Along with the grass the dust should be greatly decreased.

McArthur FFA Takes The Trophy In Livestock Judging. In 1959 the Fall River High School boys captured the blue ribbon in the livestock judging contest. Bill Estes placed 1st individual with Ernest Bruce placing 3rd individual. David Bruce and Melvin Crum tied for 4th, Bill Kelly placed 5th, Joe Bruce placed 7th and Donnie Crum 8th. The judging team composed of Bill Crum, Art Boster and Clayton Oilar judged for practice but their scores couldn't be counted in the team scores.

1965 was the first time the fair was sanctioned by the American Quarter Horse Association and the Pacific Coast Quarter Horse Association. Many new activities have been added to the 1965 roster of events, including a destruction derby and an evening fireworks display.

Many new buildings have been constructed on the fair grounds since the series of Quonset buildings. In 1949 a beef barn, rest rooms and the large community building and cafeteria were built. In 1951 the sheep and swine barn was added. 1952 was a building year with the Agriculture building, Judging building and horse barn added. 1953 to 1959 the weight and scales building, bleachers, grandstand, concession building, ticket booths, floriculture building, commercial building and the livestock sales building were added to an expanding fairground facility.

In 1962 the office building was constructed which also had office space for the farm advisors. A caretakers residence was added in 1964. The Jr. Livestock barn was added in 1967, a junior exhibit building in 1970 and the arts and crafts building in 1977. Yearly remodeling and repairs are always on the agenda to keep the buildings and grounds in good condition.

FIFTY GOLDEN YEARS OF THE INTER-MOUNTAIN FAIR CELEBRATED IN 1968.

With the sun shining and the flowers blooming it was fair time again in McArthur, California. The hustle and bustle was as it had been in the past years. The carnival getting set up, the last minute repairs. Livestock being unloaded, pigs squealing, lambs bleating, unsure of their new surroundings. Commercial exhibitors setting up their booths and feature exhibits displayed for the enjoyment of the fair spectators.

The biggest flowers and best floral arrangements put into place at the last moment so they will be fresh. Tasty pies, cookies and cakes ready for the judge to sample. Jars of fruit, vegetables, jams and pickles displayed also for the judges approval.

Beautiful quilts, clothing, sweaters, and fancy work to be evaluated for that coveted blue ribbon. The flower borders groomed to perfection, the biggest pumpkin and shiniest apple, all ready for the approval of the judges. Yes, it is time for the fair to begin.

But this year it is a special fair. 50 years of hard work has gone into the success of the Inter-Mountain Fair. A lot of volunteers are to be commended along with George Ingram, fair manager for such a magnificent fair.

The shiny exhibit halls, the flashy parade, the rodeo and other shows, and nearly 8,000 exhibits are possibly the biggest tribute to those organizers of that first fair held in mid-October 50 years ago.

Friday night there will be 4-H and FFA showmanship contests and the carnival will open its gates to fun seekers.

Fair activities get into high gear Saturday with livestock being judged. Children throughout the area are invited to bring their pets to the annual Pet Show where prizes and games will be provided for participants.

Roping competition will be held throughout the day at the rodeo arena and beginning at 8 P.M. the loggers will begin competition in their annual show. A teen age dance will also be held Friday evening.

On Sunday the annual fair parade will leave Fall River High School for its march through town to the fairgrounds. Those interested in participating can sign up at the parade.

The annual Dick Hemstead rodeo, with an outstanding string of bucking horses will be featured in the afternoon.

One of the biggest attractions of the fair each year, the Destruction Derby, where cars get smashed into huge piles of junk, gets under way Sunday evening.

Monday morning activities get under way with the first annual Junior Livestock Sale. About one hundred 4-H and FFA members will offer their market animal for sale to the public.

In the afternoon, couples who have celebrated their 50th wedding anniversary will be honored at a dinner, followed by the Old Timers program. Many of those who participated in the first fair and rodeo will be present. The annual horse show will also be performing in the rodeo arena.

The size of the fair and participation has come a

long way from that first fair in 1919. The first fair was described as a success, despite "slight mistakes incidental to a totally new enterprise."

Dignitaries at the fair to help celebrate the Golden occasion were Asa Doty, Mr. and Mrs. Parker Talbot, Mr. and Mrs. George Farmer, (a bronc rider in the first fair), A.H. Hippy Burmister a cowboy hall of fame bronc rider, Mr. and Mrs. Bill Allen, one of the early bareback riders, Montana Red (Mr. and Mrs. Don Tate) one of the early bronc riders, Mr. and Mrs. Ned McGrue, Dick McGrue, Mr. and Mrs. Roy Farmer. All of these men were rodeo competitors in the early day rodeos of the Inter-Mountain Fair. A rough and rugged life of the rodeo riders, but they love the sport and were a part of history of the Inter-Mountain Fair.

Gearing up for the 1969 Fair all went well and 20,000 people poured through the gates to get in on the excitement. The Destruction Derby roared, banged and crashed in front of a full house. Even more spectators were on hand Sunday for the annual rodeo. There were 65 entries in the parade that wound its way through the streets of McArthur.

First place awards for the wide variety of booths at the fair include Inter-Mountain Artists in the Specialty Booth, McArthur Grange in the Variety Booth, Daryl Hawes in the One Family Farm Booth, Susanville FFA in the Agriculture Booth, Cloverleaf 4-H in the Agriculture Booth and Hat Creek 4-H in the Home Ec. Booth. First place trophies awarded for parade entries include Fall River Chamber of Commerce for organizational floats; Roscoe Clark DeMolay for Junior organizational float; Thunderbird Drill Team, mounted group; Lassen County Search and Rescue for commercial entry; Fall River High School marching band; Hat Creek 4-H for novelty group; Earnest Horn, best old vehicle; Rudy Truesdale, horse drawn vehicle; Mike McCullough, single novelty; Young and Dreuzed, best matched pair; Nancy Anglemeyer, best dressed Spanish woman; Patty Griffith, best dressed Spanish Jr.; and Terrie Peterson, working cowgirl.

1968. Old timers that took a very active part in the early rodeos and fairs. Back Row L. to R.: Asa Doty, Fair Director; Parker Talbot, Farm Advisor; Bill Allen, Bareback Rider; Mrs. Parker Talbot; George Farmer, Bronc Rider; Mrs. George Farmer; A.H. "Hippy" Burmister, Bronc Rider; Montana Red (Don Tate); Ned McGrue; Dick McGrue. Front Row, L. to R.: Mrs. Bill Allen; Mrs. Roy Farmer; Mrs. Don Tate; Mrs. Ned McGrue; Roy Farmer.

Birdie Bidwell and her daughter Marjorie Earnest manning the Inter-Mountain CowBelles fair booth.

Flower arrangements are a popular competition.

Get ready for the carnival.

The flower borders contest along the walkway for all to enjoy.

In 1980 the McArthur Fair was declared a department of Shasta County by the board of supervisors. This was largely a formality as the county already owns the fairgrounds and the improvements and they approve the fair's budget. Shasta county pays most of the three full time employees wages and benefits.

In 1987 Draft horse categories have been added to the horse show in response to increased community interest in the big animals. In the early years the draft horse competition was a very important part of the fair. At that time draft horses were used in every aspect of agriculture.

A unique fashion show was sponsored by the Inter-Mountain Cattlewomen at the 1987 fair. One hundred years of fashion from 1840 to 1940 were the clothes that were modeled. Wedding dresses from the late 1800's, riding skirts, Camp Fire Girls uniform from the early 1900's and many beautiful gowns and clothes were shown. Children's clothes and men's apparel were also modeled. The show was produced in front of a full house. Many families gathered to see Grandma's wedding dress displayed and as the dress was being modeled, the bride of many years ago stood for recognition and applause.

In 1992 it became necessary to charge a gate fee for the entrance to the fair. The board voted and passed the motion that needed to be done to defray the expenses of the fair. On Friday is Senior Citizens Day and they gain free entrance to the fair and on Saturday is Kids day and all kids get a free day at the fair. Monday there is no gate charge for anyone.

Also in 1992 a very unique gazebo was added to the landscape of the fairgrounds. Bob Wimer wanted to honor his wife Lela and offered to build and dedicate a gazebo on the fairgrounds to her. A gift to the fairgrounds and is a beautiful setting for weddings, concerts and stage productions. Lela's Gazebo is a welcome addition.

The annual events at the fair are always the Horse show, Golden Wedding Dinner, Pee-Wee Swine Showmanship, Slide Show, Pet Show and the Diaper Derby, Ladies Lead, Scout's Pancake Breakfast, Junior Rodeo, Parade, Destruction Derby and the Junior Livestock Auction. There is always a dance and entertainment and demonstrations constantly to keep everyone's enjoyment. The carnival and exhibit buildings are open daily and there are so many food concessions you could not possibly try them all.

The 76th Annual Fair with a theme of "Ribbons and Bows and Cowboy Clothes", closed its gates on the 1994 event with a huge success. Each and every year is better and better with the "Showtime In The Mountains", 77th annual fair having 11,777 exhibits shown by 3,433 exhibitors that participated in making this one of the best fairs ever for 1995.

Hugh Carpenter 1954
Horse Show
pulling contest.

George and
Olive Brown
Grand
Marshall's
of the
parade in
1966 with
queen
Janet Oilar.

Kristy and
Kerry Klagues
with their
4-H projects
that they
entered in
the fair
in 1977.

Judy Perkins, Pittville Raiders practicing to ride in the parade.

Lori and Bryan Gerig having a little jam with 1966 fair queen Janet Oilar.

ABOVE:

1971 Queen Lani Bickle and Mike McCullough trick roper.

AT LEFT:
Trick rider and roper Mike McCullough.

It was a great fair.

Harry Flournoy, announcer,
Bonnie Sattler, Clerk and
Pat Huff having a good laugh.

A tornado took the roof off of the hog barn and slammed it into the quonset building the day after the 1962 fair.

The tornado smashed part of one of the buildings into the barn where Charles Bruce was shearing his sheep.

The Busy Fair.

A bustling Inter-Mountain Fair.

1977. The new art building under construction.

Office crew: Kristi Johnson, Valerie Lakey, Kris Cunningham, Carol Sanders.

The Scarecrow Contest.

Artists in Action, Argent Hale.

Those flowers that Rose planted.

Marshall Flowers on Snake Eyes.

Snake Eyes

It is Indian Summer 1938. The blue haze of early fall veiled the snowy Sierra peaks that hem in the beautiful Fall River Valley. Mount Shasta, king of the Cascade Range, dominated the view to the northwest, while the queen of the Sierra, Mount Lassen kept watch over the valley from the south.

It was fair and roundup time for McArthur. Horses and men of every known pedigree (and some that weren't known) had invaded this mountain cow town. Livestock, culinary achievements, and handiwork had all been judged and the rodeo was attracting the main attention. Bill Lee, a local cattleman and champion horseman in his own right, was ramroding this western spectacle. Raised in this northwestern country on a large cow spread, Bill had the nature and the background to cope with horses and men that followed the rodeo sport. Then, too, his handsome appearance and pleasing personality coupled with a robust physique and a full knowledge of rough and tumble fighting were fine qualifications for a director of this western drama. In his prime no cowboy ever crowded him twice. His reputation was known far and wide as a fair and competent arena boss.

In those democratic days, horses used to try the contestants' skill were assembled from nearby ranges and ranches. No professional buckers were used. All animals were rounded up a few days before the show and were given numbers and names depicting their agility, color and disposition, Pit River, Juniper Jim, Yellow Fever, Black Bart, Bloody Island, Hounds of Hell, Swamp Angel and Snake Eyes were some of the tough buckers back in those "new deal" thirties.

Bronc riders of note and fame were Perry Ivory of Alturas, Jack Meyers of Red Bluff, Bob Lockee from the parched lands of Arizona, McKinley Machach, a native Paiute and last but not least, Marshall Flowers from Cottonwood, the poison oak district of southern Shasta. All were good men with spur and saddle.

Riding in the finals for top money on this dusty, balmy Sunday afternoon were Flowers, Meyers, and Lockee. Judging this tough, fast moving event were Bill Eldridge of Pittville, Bert Jensen of Susanville and Roy Owens of Red Bluff. At this stage of the show all attention was focused on Flowers and a horse called Snake Eyes.

Flowers was a handsome, black haired twister, standing over six feet tall and weighing about 230 pounds. It was said that with a little luck he could contest any horse on the circuit.

Snake Eyes was a blood bay, slim built clean limbed gelding. His head was large and flat with small ugly eyes resembling those of a rattlesnake. A 44 brand was stamped on his left hip and a CI iron was burned deep on his left shoulder, indicating that he had more than one owner. This horse had been raised in the desert sagebrush country of Winnemucca, Nevada. His conformation indicated that a large amount of Standard bred blood flowed in his veins. Fred Andrews, a horse trader, had brought Snake Eyes to the Fall River country and sold him to W.J. Albaugh a prominent Pit River rancher. Albaugh had attempted to break this horse to the harness, but he proved to be a kicker, the reason that Bill Lee had him in his bucking string. Snake Eyes had no reputation as a bronc but Lee reasoned that any kicking horse was a bucking horse. Marshall Flowers drew this horse to compete in the finals, as he was ahead on points, any average ride would bring him the coveted first money.

As Snake Eyes was driven into the bucking chute he started a kicking spree. Marshall grabbed his panther tracked scarred saddle from the arena dust and tossed it up over the chute. The horse kicked even harder. Somebody eared the horse down and another cowboy screwed on Marshall's "chair". The flank rigging was put in place. While this was going on, Marshall was putting on his chaps and spurs in a measured precise manner. Lee looked over the bucking chute and ordered an attendant to pull the flank rigging up one more notch. Marshall settled down on the summit of Snake Eyes, pulled his hat down tight, took a deep seat in the old association saddle and said, "Let me have him". The gate flew open and this Nevada raised bronc came out of the chute like a scalded cat. He leaped high in the air and kicked high behind, in fact, so high that he almost tipped over forward. Marshall was spurring high in the shoulders at the same time waving at the grandstand. The next jump Snake Eyes swapped ends and the third jump he repeated this maneuver in the opposite direction, that's when the big burly cowboy from Cottonwood and Snake Eyes parted company. Marshall bit the dust, a disappointed, angry cowboy. The tense crowd roared! They wanted Jack Meyers to win the money, which he did by making a spectacular ride on Yellow Fever, a bronc from the Little Valley country.

The show ended, the chutes were emptied and the alkali laden dust in the arena settled to the ground. Dusty, sweaty cowboys as well as cheerful spectators made their way into the Buckhorn, a corner saloon known as the "water hole" by the natives. As Marshall Flowers passed though the pubs open door where songs and witty stories crept through, he elbowed his way quickly to the bar and snapped, "Give me a double whiskey three fingers high on the rocks." Bill Albaugh, owner of Snake Eyes, said to Marshall, "Why don't you mix some water with that whiskey, drink it slow and easy and enjoy your drink?" Marshall answered by saying, "When I drink whiskey I want to feel the effects of it right now!"

Continuing, Bill said, "You had hard luck this afternoon, I guess you found that old bay horse of mine kinda hard to ride."

Marshall, in his gruff way replied belligerently, "Maybe you think I can't ride that ole horse." Bill answered, "Yes, Marshall, I think you can if you'd hang on a little!"

The gay crowd laughed!! Marshall gulped down another whiskey and turned to face the crowd. Hitting the bar with his huge fist he growled, "I can ride any horse in them there corrals, contest style."

No one in that smoke filled room, steeped with the odor of thirst quenching liquor questioned Marshall's bronc riding ability. So on that Autumn day in that little western town he was loser in the arena, but champion at the bar.

Written by Reuben Albaugh and added to this book by permission of Albert Albaugh.

Events of the Fair

PET PARADE

The children's pet parade is exciting and fun for all ages. No matter what your pet may be, a fat green frog, the fluffiest kitten or the biggest dog ever, you can bring it to the Pet Parade.

With your pet in your arm, tagging along behind or pulling a wagon you can be in the pet parade. Dressed in your best costume with your pet in his or just a ribbon around its neck, what ever, you can enter in the parade and win a ribbon.

The funniest pet, prettiest cat, most unusual pet, dog with the waggliest tail and smallest child with biggest pet are just some of the categories that are on the roster. So come on down and have some fun.

Those in 1967 that had winners were Best Dressed Dog, Kim and Shane Estes with Gidget; Biggest Dog, Alice Ketman, with Shaffer; smallest, Gail Rogerson with Peppy; most tricks, Allen Lung with Spike; loudest bark, Chris Carpenter and Tramp.

In the cat category, Best Dressed, Nancy Eldridge and Teddy; biggest Cathy Carpenter and Penny; most colorful, Margo Eldridge and Pebble; shortest tail, Connie Cessna with Paula; purringest, Norman Schneider with Midnight. The most unusual pet was a caterpillar shown by Ronnie Rhodes. Best Costume went to little Mike McCullough with a suit resembling Uncle Sam and his dressed up dog.

Mrs. David Schneider and Mrs. Merrill Cessna were in charge of the event with Al Pratt as Master of Ceremonies. Asa Schneider was in charge of the music and David Schneider handed out the prizes. Game winners were Billie Cessna, Rena Oilar, George McArthur and Joyce Rhodes.

Pet Parade winners Libby and Maggie Decoito.

WATERMELON EATING CONTEST

Winners were Donnie Estes, Dean Albaugh, Lennie Oilar, in group 1. In group 2, Jimmy Estes, Wanda Hensall and Danny Dabill. Group 3 winners were Dan Cook, Jim Arnold and Rod McArthur.

PUMPKIN CONTEST

Each year many children ages 6 to 12 who live in the Fall River and Big Valley school districts pretend they are Charlie Brown and spent their summers tending pumpkin patches.

Prizes are given for the largest pumpkin and the pumpkins with the funniest, scariest, and prettiest faces.

Jeff Schneider and his giant 1st place pumpkin.

This event must be registered for in June, so the children have time to cultivate and grow their pumpkins and care for the plants themselves. The prizes and ribbons are well worth the effort and fun.

BLUE RIBBON CITIZENS

They are people of the community that go that extra mile to do their part to make sure the fair is a better place each year. Asking nothing in return, sometimes not even a thank you, but they still are there to do their job as volunteers doing something they believe in and love. Each year after the fair is over the board agree on a person or organization that has done the fair a great service above and beyond the call of duty. Paid employees are not eligible for this award. The award is made available annually through the Western Fairs Association.

AT RIGHT: Catherine Ryan Blue Ribbon Winner and George Ingram, 1981.

1971. George Ingram, Delores Leone with her Blue Ribbon Award and County Supervisor Norman Waggoner.

The Blue ribbon award was established in 1966 and recipients of that commendation are; Everett Beck, Sam Thurber, Lawrence Agee, Willis Albaugh, Rose Schneider, Frances Gassaway, Delores Leone, Dick Nemanic, Burney Lions Club, Aaron Stockton, John Boyes, Vern Cunningham, Thomas Vestal, Norma Callison, Catherine Ryan, Fall River Big Valley Cattlemens Association, Frank Parker, Ron Taylor, John and Shirley McArthur, Kathi Corder, Bank Of America, Norman and Betty Taylor, McArthur Volunteer Fire Department, George Ingram, Wally Hilliard, Ed Bruce, Gail Ashe, Rose Schneider (second time) and Inter-Mountain Conservation Camp.

LADIES LEAD

Bring your fanciest and fittest sheep and wear your best wool outfit and you are ready to join the parade of the ladies lead. Established in 1960 and a grand idea of Susan Taylor, this popular event is a favorite.

Entries in this class will be open to any young ladies over the age of nine. Sheep must be conditioned, fitted and trained to show at halter. Exhibitor need not be the owner. No rams will be allowed. The entrant will be judged on the mode of dress selected. Clothing may range from a scarf to a three piece outfit, with material being at least 70% wool and the articles need not be

ABOVE:

Aimee Thompson with her sheep in the Ladies Lead Contest.

AT LEFT:

All ready for the ladies lead.

1975 Destruction Derby.

homemade. The outfit the sheep wears matches the one of his mistress as they parade in front of an enthusiastic audience.

A Frog Jumping contest and Watermelon eating race is also a part of the fair for fun time activities. In 1967 Bruce Arnold won the frog jumping contest with Freddy The Frog when he jumped 102 inches.

THE DESTRUCTION DERBY

The name of the game is smash and run in the ever popular annual Destruction Derby. This event started in 1969 and is looked forward to each year by many.

This exciting event is sponsored by the Burney Lions Club at the fair grounds arena. This smashing good time is usually viewed by a sell out crowd with standing room only.

A four wheel drive truck pull made it's first appearance at the fair in 1987. This event shows the brute strength of the most powerful truck with the most proficient driver. A crowd pleasing event does not make an appearance ever year but is a special competition.

GOLDEN ANNIVERSARY CELEBRATION

Being married 50 Golden years is something to celebrate, and the Inter-Mountain Fair is a part of that occasion.

Each year the Fair sponsors a special dinner in honor of those couples that have reached that half century of marriage. In the beginning the celebration was called Pioneer Day, but Willis Albaugh thought it was a good idea to honor those that have reached a special milestone in their lives.

Frances Gassaway was the first hostess. For many years Norma Callison made the wedding cake then fell into the job as hostess for ten years. Norma sent out fifty invitations and that amount has tripled by now. Aldora Hornbuckle was the next person responsible for being the hostess and now Alexis Johnson has that responsibility.

A full turkey dinner with all the trimmings is usually on the menu, with a four tier wedding cake being served for dessert, all courtesy of the fair.

Entertainment is provided along with door prizes

given. It is a very enjoyable afternoon for the several hundred senior citizens that have been invited to attend. This is probably one of the events that is looked forward to more than any other for those charming folks.

BUCKING LOGS FOR THE FUN OF IT

Logging changes its mood from work to play when tree fallers come to test their skills at the Inter-Mountain Fair. The annual logging show provides entertainment and pleasure for all those who appreciate woodsmen's skill.

Contestants participate for both prize money and trophies in the following events; Power Saw, Chopping, Axe Throwing, (for men, women and juniors), Hand Bucking, (double and single), Steeple and Jack and Jill.

Women's Axe throwing is open to all women, but competition in all the other events is limited to residents of Shasta, Lassen, Modoc, and Siskiyou counties.

To prepare for some of these events, special tools and the skill to use them are necessary. The power saw event is actually called One-Man Power Saw, Direct Drive. In this event, each contestant will make one cut, bore through, and cut up or down until the cut is completed, all against time. In the chopping event an axe is used to chop through a log about eight inches in diameter. The Axe Throwing event requires an axe and a target similar to a dart board. Each contestant will make three throws with the highest score winning.

The hand bucking event requires the use of a cross cut saw known as the Peg and Raker. These saws are hard to come by, because they are no longer made. The logger must cut a ten inch log. The contest includes single and double and Jack and Jill, a man and woman team.

The Steeple Chase is a timed event. Entrants must carry a gas can, axe, wedge, and power saw from the starting line to the first five poles. They set their equipment down, start the saw, saw off the cut at the top of the pole, shut off the saw, pick up all the equipment, and move to the next pole. They repeat this procedure at all five poles. Time ends when the contestant crosses the finish line with all the equipment.

A crowd pleasing event, the logging show is very popular in this country where the timber is one of the main industries.

Vern Cunningham in the axe throwing competition of the Logging Show.

Ardis and Vern Cunningham
in the Jack and Jill contest
of the Logging Show.

Novelty axe chopping in
the Logging Show.

AT LEFT:

1980. All Around Logger Fred Schneider.

BELOW:

Scott Herringer oiling for Buzzy
and Lennie Eades in
the Logging Show.
Buzzy, All Around Logger, 1967.

Fred Schneider, All Around Logger and Queen Cindy Campbell, 1980.
Kris Cunningham, clerk, and Harry Flournoy, announcer.

Peggy Perkins and her steer, 1967.

Carrie Stone, granddaughter of Catherine Ryan getting her bull ready to show in 1976.

Livestock:
the Hustle and Bustle of the Fair

THE JUNIOR
LIVESTOCK SALE

Each and every year the barns are bustling with activity with the members of the FFA and 4-H Clubs getting their animals ready to show and sell. On Friday the showmanship and market classes take place with scrubbed and groomed animals and kids.

Everyone is trying hard for that Showmanship Trophy or the coveted Championship Ribbon and the number one animal. Members from Fall River, Burney and Big Valley are eligible to show and sell at the Junior Livestock Auction that is held the last day of the fair Monday.

Labor Day 1966, generous buyers turned the first annual Inter-Mountain Junior Livestock Sale into a huge success and they paid the highest average price for animals seen at a 4-H and FFA sale an any Northern California county fair this year. Without a flinch buyers paid exceptionally good prices for the sixty-six animals

1993 Junior Livestock Sale, Dick Nemanic, Kathy Lakey, Wade McIntosh, Sue Hoffman, Elena Albaugh, (in front) Tina Taylor and Hugh Wilson.

105

Betty Bruce and her Southdown Lamb, 1960.

sold. Cynthia Pratt's grand champion swine was purchased by Earnest and O'Neil Hereford Ranch for .75 cents a pound. Jeanne Lakey received .78 cents a pound for her grand champion beef from John McArthur Farm Supply and Vicki Thompson's grand champion lamb was purchased by Shasta Livestock for $2.00 a pound.

Joanna Doty placed first in the Showmanship contest held for the 4-H and FFA. Miss Doty won the award in the beef cattle with her black Angus steer. Peggy Perkins also placed in the showmanship with her steer. FFA members that placed in the showmanship were Robert Thompson of Hat Creek and Danny Bouse of Bieber. Those local members placing in the sheep division were Tracy Nemanic, Leo Estes, Hank Rodman and Henry Crane. Swine showmanship winners were Tom Wayman of Bieber, and Chuck DeLapp of McArthur. Judges in this competition were Jim Earnest and Donnie Crum, both past members of the McArthur Chapter of FFA and 4-H. This was the first year local young men were accepted for this position.

Dick Nemanic sale chairman was happy with the outcome of the sale. There had been some doubts about starting the sale at McArthur and fears that it may not be a success. Parents were hard pressed to take their children and animals to the fair at Anderson as the expense and time away from home was a difficult situation.

Charter buyers that bought at that first sale and still are active in the sale today are, Ernest and O'Neil Herefords, John and Shirley McArthur, Shasta Livestock, Albert Albaugh, Bank of America, Fall River Feed, Federal Land Bank of Red Bluff, McArthur Electric, Packway Materials, Shasta County Farm Bureau, Tri-Counties Bank and Tom and Marie Vestal.

A special feature at this auction since 1968 has been the sale of the animal of which the proceeds will be donated to the Pete Lakey Memorial fund. Each year a hog has been donated by the Stoltenberg family along with the feed to take care of the animal. The animal is raised by a member of one of the 4-H clubs or FFA and sold at the auction for a benefit. The money is used for medical, welfare and educational assistance to 4-H and FFA members in

the area. Pete Lakey was the son of Asa and Mary Lakey and died of cancer before he could sell his animal at that years fair. His pig was auctioned off and thus the start of the fund.

George Ingram said that when the idea of a Junior Livestock Sale was proposed to the community there was a lot of negative comment. It was a struggle to get it going, but when it finally took off, it has grown by leaps and bounds.

Dick Nemanic has been the chairman of the sale board as long as it has been going and is credited with a lot of its success. Shirley McArthur was named by Dick as one of his most valuable helpers. Many volunteers are credited for the prosperity of the sale.

From a handful of animals in 1966 it has grown to 137 Market Hogs, 44 Sheep and 31 Steers in 1995. The community support is overwhelming for this activity that was a struggle to get started.

Willing to pay the higher prices, the community has accepted the auction in order to help the youth of the community.

Each year a buyers dinner is held for the people

Mark Bidwell and his first Grand Champion Steer in 1969.

1973. Stan Weigand buyer of the Grand Champion Hog of Phil Watkins.

Kris Cunningham and Wally Tyler.

who have purchased an animal at the Junior Livestock sale. This is a way of saying thank you to the people who have made this sale possible by purchasing animals. An award is given to those repeat buyers and the top buyers of the sale. The 4-H and FFA members participate in helping with the dinner.

In 1967 competition was exciting for the 4-H and FFA members. The youngsters competed in the showmanship contest in record numbers. In the FFA Showmanship division, first in beef went to Norman Newell of Cottonwood, first in swine to Pete Lakey, and first in sheep to Frank Watkins.

In 4-H Peggy Perkins walked off with first in beef while Jean Hall took first in swine. Susan Crane took first in the sheep division.

Every 4-H and FFA member that enter livestock at the fair hope for that coveted championship ribbon. Some are lucky enough and work hard enough to receive the judges nod several different years. Perry Thompson from Hat Creek was one of those lucky hard working ones and had about four champion steers and won the Showmanship contest three or four times.

Wally Tyler was always a familiar face among the hustle and bustle of the Junior Livestock barns. The 4-H Youth Advisor for the University of California Cooperative Extension works intensely on making sure all the animals, participants, and personnel are where they should be during the Junior Livestock Sale. Wally knew it could be heartbreaking for some of the kids to not make weight with their animals or be sifted for one reason or another. Or if they did make weight and were able to sell, there were always tears from some to have to give up that animal that has been their friend and companion for several months. Norman Taylor and Ernest Bruce were there to weigh the animals and hold their breath when the scales just barely tipped on the right mark. Watching that wide-eyed child that is putting all their faith in the weigh master to say things are O.K. An anxious child waiting for him to put a sale tag in their animal's ear, the little numbered tag that is a seal of approval.

The 1974 Junior Livestock Sale had 24 hogs enter the ring. Vicki McArthur had the Grand Champion Swine and sold it for $1.65 a pound to Central California Federal Service. Glenda Edwards 1,135 pound beef wore the purple ribbon of the Grand Champion and sold to Shasta County Bank for .95 cents a pound. Thirty-five beef were sold through the auction ring. Ron Johnson took top honors with his 100 pound lamb selling to Federal Land Bank of Red Bluff for $2.50 a pound. Sixty lambs were sold to make a total of 119 animals to go through the auction ring. The average

1974. Norma Oilar and her FFA steer.

Wes Lovrin
and his
FFA hog,
1977.

Mark Bidwell and his last Grand Champion Steer in 1977.

price paid was .92 cents for swine, .71 cents for steers and $1.27 for lambs.

In 1977 Grand Champion junior market animals were selected by judges from the largest field of animals ever shown at the fair. Mark Bidwell's steer got the nod from the judges for the top honors. The top steer was purchased by Federal Land Bank of Red Bluff for .76 cents a pound. The Hi Mont Motel paid Lynn Marsters $2.85 a pound for her Champion lamb and the Federal Land Bank also purchased Jean Halls Grand Champion hog for .76 cents. Prices at the livestock sale were good as buyers brought the average price of steers up to .46 cents a pound while paying an average of .99 cents for lambs and .70 cents for hogs.

The 1988 Junior Livestock Sales were up that year with a mixture of smiles and tears as warm, loved animals were exchanged for cold, hard cash. The beef showed an average of $1.63 a pound with the sheep bringing an average of $5.91 a pound and the hogs brought $3.13 a pound. The Pete Lakey Hog sold fifteen times for a total of $6,545.00.

1995 was a good year for the 4-H and FFA students when 213 animals were sold through the auction ring. Members stashing away their checks to go to college, a new bicycle or to purchase another animal for next years fair. Twenty-nine steers were sold and Qwen Lakey had the Grand Champion with Tanner Songer leading the Reserve Champion. One hundred and forty-six lambs were auctioned off and James Estes had Grand Champion and Carson Estes had Reserve. Kimberly Bernard showed the Grand Champion Swine and Mike France had the Reserve and a total of one hundred and thirty-six hogs crossed the auction block.

In the beginning of the Junior Livestock sale, Aaron Stockton of the Orland Auction Yard was the auctioneer, with his auctioneer chatter getting the highest bids for the 4-H and FFA members. For many years the crew from the Shasta Livestock Auction yard have done the honors and work their hardest to get the highest bids possible. The fair and the owners of the animals are very grateful for their expert help.

Heather Dye, 1987.

111

Dawson Hartman, 1985.

Cheryl Bruce and her 1994 Champion 4-H hog.

Showmanship winners, 1995. Katie Bosworth, Jennie Dye, Tori DeBraga, Jason Hoffman and Emily Kelly.

Mindy Haury and her hog, 1995.

Russ Carpenter and his hog, 1995.

Trina Albaugh and her steer, 1995.

Baley Thomas and her
FFA Swine Project.

Lucas Stevenson
and one of his prize rabbits.

THE STEER FUTURITY

The steer futurity developed from the pen show that had been a part of the Inter-Mountain Fair for many years. In 1987 there were 13 pens of mountain grown steers competing for the number one slot and labeled as the best. In 1991 the pen show had grown to 99 pens and everyone was excited to enter their 650 to 999 pound steers. Not only are the steers judged as part of the livestock show at the fair, but they go on to higher evaluation as soon as the fair is over. Each pen of five steers are shipped to a feedlot where they are evaluated for rate of gain and feedlot performance. The carcass evaluation is the final phase of the competition. Awards

are given in rate of gain, feedlot performance and carcass evaluation. The top four of the five steers are judged in this division. There are no boundaries in the pen show and it is open to the world.

The steer futurity is a tool in letting the producer know what kind of performance their animals can accomplish in a controlled situation.

The Champion pens of cattle at the pen show were in 1957, Albert Albaugh; 1958, Kenneth McArthur; 1959, Albert Albaugh; 1960, Bill Kelly; 1961, Ronald Hutchings; 1962, '63, James Albaugh; 1964, Harley Neuerburg; 1965, '66, '67, '68, '69, Albert Albaugh; 1970, Floyd Bidwell; 1971, '72, Harley Neuerburg;

Working the Pen Show, Kurt Urricelqui, Joanne Bruce, Ruth Taylor.

1973, George McArthur; 1974, Lem Earnest; 1975, David Gates; 1976, Rod McArthur; 1977, '78, George McArthur; 1979, Stephen Albaugh; 1980, George and Rod McArthur; 1981, Stephen Albaugh; 1982, '83, '84, Tom Vestal; 1985 George and Rod McArthur; 1986, '87, '88, '89, Tom Vestal; 1990, Norman Taylor; 1991, Bidwell Ranch; 1992, '93, Little Shasta Ranch; 1994, Jerry Hemstead; 1995, Little Shasta Ranch.

The pen show is co-sponsored with the Fall River-Big Valley Cattlemen along with the Inter-Mountain Fair. And they are proud to have the second largest pen show in the nation, and the world, second only to Denver, Colorado.

The 1995 pen show had thirty exhibitors and around 250 head of cattle. This division of the fair has had as many as 400 head entered with 95 head going to the futurity.

FALL RIVER-BIG VALLEY CATTLEMENS FEEDER SALE

Each year in October the local Cattlemen band together to have a feeder sale to market their animals. On October 10, 1958 the first feeder sale was held to market local cattle. John McArthur was chairman and Albert Albaugh, Lem Earnest, Hardy Vestal, Andy Lakey, George Brown Jr. and Walter Callison all worked together to get a new enterprise going. Margie Earnest was the secretary and did all of the books for the feeder sale.

The Cattlemen banned together and decided it was a good thing to try to have a livestock sale to market their cattle without having to haul them for miles to the nearest auction yard. In the beginning they tried to have a spring and fall sale, but found there was not enough cattle ready in the spring to warrant that one, so they dropped it. The Fall sale has proved successful and around 1500 head are marketed annually. The feeder sale brings top dollar and with a good buyer attendance it is a popular place to sell your cattle and get the top dollar.

The Inter-Mountain CattleWomen serve a noon beef dinner that is hard to beat, doing their share to help out the cattle industry and promote beef.

Maybe the forerunner of the feeder sale was in 1957 when the Fall River Ag teacher, Ron Hutchins asked some of the cattlemen to help him sponsor a FFA fancy feeder sale. The sale was held for two years but was not successful.

The fancy feeder calf sale as it was called was

Feeder Sale 1970. L. to R. (on ground) Pete Lorenzen, Albert Albaugh, Rob McGregor, Jerry Parks. (On stage), Bobby Jones (Auctioneer), Lyle Christensen, Willis Albaugh, Ernest Bruce, Carl Carnahan, Walt Johnson, Ellington Peek.

117

1st Feeder sale in 1958.
L. Lem Earnest, R. John and
Kenneth McArthur.

sponsored by the McArthur FFA and resulted in 97 calves being sold at an average of 38 cents per pound.

The highest price registered at the special sale was 52 cents per pound paid by Howard Hayward of Overton, Nev., to Morris Doty of Cassel, for a 455 pound Angus steer. McArthur Brothers of McArthur sold the calf which brought the second highest price, a 485 pound shorthorn which Hayward also bought for 51 cents a pound. The top 10 averaged 50 cents per pound. A 420 pound Bidwell calf sold for 47 cents and a 520 pound calf from the McArthur Brothers sold for 48 cents. The first 100 calves through the ring averaged 41 cents, but the total sale averaged 38 cents on the 126 calves sold.

The special calf sale followed the sale of the Fall River Cattlemens Association in which 1,622 head of feeder cattle sold for an average of $162.09, slightly less than last year.

In 1965 Floyd Bidwell was the president of the Fall River-Big Valley Cattlemens Association. This active group had 102 members who own some 17,000 cows. The feeder sales has brought the association a lot closer together. John McArthur said this was one of the side benefits of the feeder sale. After the lower cattle prices in the 1950s it forced cattlemen into action. That is what led the local cattlemen into creating their own

feeder sale in 1958. The fair grounds were available and a perfect place to hold the feeder sales. Sam Thurber said there were quite a few fifty-head outfits here and marketing has always been a problem for them. Thurber, the local farm advisor was very instrumental in helping get the sale going. Most local sales for small outfits had primarily been to the low caliber order buyer who bought their cattle for his price, take it or leave it. The rancher ended up taking it.

But in 1958 when the cattlemens association feeder sale committee launched the group's first home spun feeder sale which has today expanded into a top notch buyer drawing, triple auction success. The local sale helped small cattlemen 100 per cent. All we had before were order buyers, now about 30 buyers attend the sale. In 1964 the steer calves were bringing 18 and 19 cents. Twenty-three cents was the highest at the sale and the average was 21.50, the ranchers felt that was pretty good.

In 1963 two sales were held, one in March and one in October. A third one had been held in November, but it was not too successful. More pens are needed to house the amount of cattle run through the local sales.

One hundred gates had been hung on a community work day sponsored by the cattlemen to finish up more corrals. The fair furnished materials and the cattlemen

Everett Beck at the feeder sale.

built the corrals to hold 2,400 head of cattle.

The 1966 sale saw 1225 feeder cattle go into the auction ring. The top price of 31.85 a pound was paid for some Hereford steers. Glen Aldridge received the second highest bid of 31.60 for 25 of his fancy Hereford steers.

In the mid 1980s, 3200 head of cattle went through the ring and was one of the top sales. This last year, 1995, 1,487 head of cattle marched over the scales and the gross sales was around $580,000. From 300 pound heifers to 1100 pound steers there was quite a variety of cattle. In a declining cattle market and so many ranches in the Inter-Mountain area not raising cattle anymore, it makes you wonder what the outcome of the local feeder sale will be.

Working the Feeder Sale, Roger Urricelqui, Al Eleshio and Tom Coe.

Action Packed Rodeos

JUNIOR RODEO

The Junior Rodeo Board incorporated in 1974. The original members to work so hard to sponsor this event was Tom Vestal, Bill McCullough, Bud Knoch, Peter Gerig, Charles Kramer, Albert Albaugh and Andy Babcock.

The Junior rodeo has always been a popular attraction for the community. A huge success, the fair Junior Rodeo is known as the best, well run Junior Rodeo with the biggest assortment of prizes for the cowboys and cowgirls. Since it began in 1974 it has grown from a playday for the local children to a successful two go rodeo with four champion all around saddles given to two cowboys and two cowgirls.

Saddle winners in the past years were: 1986, Hardy Vestal, Kim Urricelqui, Shelby Pierce and Kathy Jo Brown. 1987, Larry Hammonds, Angie Wilson, Rondo O'Connor, Seth Britten. 1988, Tami Hill, Steve Stone, Ben Britten, Sara Livingston. 1989, Rondo O'Connor, Cory Thomas tied with Amy Livingston, Wyatt Harbert. 1990 Rondo O'Connor, Shannon Brown, Seth Britten, Katie McCullough. 1992 Weston Hutchings, Kristi Cox, Luena Harbert. 1994 saw Jennifer Lemke winning a saddle and 1995 Jake Hayden won one. Several more were saddle winners, but our records are incomplete.

High school champions from the Inter-Mountain area were Coupar Thomas, Rondo O'Connor and Ben Britten. In 1989 Gina O'Connor went to the nationals and in 1988 was Cow Palace all around. State Cham-

1966. Charles Kramer and Peter Gerig on the run.

121

I'm gonna ride this calf.

pion in barrels was Cory Thomas and Jenny Britten was State Champion in Goat Tying and in 1985 Kim Urricelqui was State All Around. In the 1993 series of gymkhanas Cowboys and Cowgirls ages 6 through 18 worked on timed events, improving their skills and that of their horse. Everyone needs the experience of Gymkhanas. The children compete for ribbons and year end points which are spent at the tack shop. As Tom DeAtley puts it everybody goes home with something.

The Inter-Mountain Junior Rodeo has been responsible for the making of many good cowboys and cowgirls. Some like Rhondo O'Connor who got his start there when he was six years old and is going on to higher places in the rodeo world. We will probably be seeing his name in the rodeo news for years to come.

Sunday is Junior Rodeo time at the Fair. In 1985 the rodeo went through many changes. In the past years the rodeo was only open to the youth in the Fall River and Big Valley school districts. Beginning this year, it is an open junior rodeo for entrants 14 to 18 years of age. Not only will this mean more competitors, as there are no residential requirements for this age group, but also it is promised that the junior rodeo will have a lot more excitement. With an open rodeo, there promises new events such as team roping, bucking horses and bucking bull events.

The local rodeo for the six to fourteen year olds will remain pretty much the same.

This is the first year that the rodeo has not been put on by the Inter-Mountain Fair The Inter-Mountain Junior Rodeo Association became a separate organization.

In 1987 Gina O'Connor reigned as the first ever Junior Rodeo Queen. A veteran of the rodeo sixteen year old Gina is an all around cowgirl, living and working on a ranch.

As an added incentive for entrants, four saddles

Do I have the goat, or does he have me?

123

Lorilee Gates around the barrels.

Jim Duncan, Angelo Leone, Sonny Cessna, Clifford Cunningham, and Harold Cessna
look on as the rider gets dumped.

donated by Big Valley Lumber, Leo S. Jones Oil Co., Pepsi, Miller, Coors, Budweiser, Ricks Lodge, H&R Logging and Fibreboard Corporation will be presented to the all around girl and all around boy in both the junior and 14 to 18 division.

HORSE SHOW

Team Penning has become one of the very popular events of the Horse Show. PTPA sanctioned penning started at the Inter-Mountain Fair in 1991. In the penning classes, a three person team works within a two minute time limit. They must cut out from the herd and pen three head of cattle with the assigned identification number. The fastest time wins. Men and women alike participate in this popular event. Winnings are paid by jackpot along with add-back money from the fair. Placings are determined by the number of entries.

Also featured in the horse show is the branding competition. The branding is done by a four person crew, two on horseback, roping the calf, one man to throw and adjust the ropes on the calf and one with the branding iron dipped in paint. Three or four calves are in the pen and after the first two cowboys rope two calves and they are paint branded the teams change positions and the others rope and their calves are branded. The team with the most calves in the allotted time wins, or finishes in the fastest time.

Cow sorting and cow horse classes are also part of the horse show.

Carson Cunningham caught the heels.

126

Bud Knoch and Lil Lambert
judging this young cowboy.

Mike McCullough on the bull and Duane Crum on the run.

Coupar Thomas at calf roping.

Cindy Campbell around the barrels.

An excited little girl with Doug Weigand holding her calf while Richard Taylor looks on.

Holly Thomas
Inter-Mountain Rodeo Queen
and M.C. Cal Hunter.

Rhondo O'Connor, 1989.

Rhondo
O'Connor,
National
High School
Finals,
1992.

Waiting our turn.

Cory Thomas around the barrels.

GENERAL RULES OF McARTHUR RODEO

---

1. Contestants and participants assume all risks to person or stock while upon the grounds, the management extending an invitation to the world, no one barred, everyone welcome, but only upon conditions stated.

2. The Association assumes no liability for injuries to contestants, their stock, or damages done their property. Owner, contestants and assistants assume all risks.

3. Substitutes will not be permitted in any event or contest.

4. The Association will select competent judges and their decision will be final.

5. Contestants or others will be ejected from the grounds and barred from all events for any one of the following offenses: **Being under the influence of intoxicants; rowdyism; quarreling with the judges or officials; or for any other reason deemed sufficient by those in authority.**

6. Management reserves the right to make any changes or additional rules as circumstances may demand in any event.

7. The management will under no circumstances tolerate cruel or inhuman treatment of stock by contestants or participants.

Everett Beck
Parade Chairman around 1960.

Asa Doty
Grand Marshall
of the parade.

Get Ready for the Parade

The Inter-Mountain Fair Parade is one of the events that is looked forward to by everyone. For a small community, the parade is something to be very proud of. For several years the parade was not always a stable event, it depended on if there were enough entries or not. By 1960 the supervision of the parade was turned over to a very willing volunteer that made sure the parade was going to roll on time. Everett Beck has been involved in the Inter-Mountain Fair Parade almost as long as he has been in the area.

In 1959 Beck moved to the Fall River area where he became the first resident California Highway Patrol Officer. A good friend of George Ingram he offered to help with the parade and for more than twenty years he very capable handled this responsibility. Beck was the very first blue ribbon winner. This is an honor that is bestowed upon a volunteer that goes beyond the call of duty to make the Fair the best there is.

Everett started planning the next years fair as soon as the last one had ended. Evaluating the last performance and making improvements for the next. Once the fairs theme is announced, three months before the fair, Beck started organizing the parade. He chose the Grand Marshalls and worked on the line up, sent invitations and started working on the entries.

Everett is proud to be a member of this community

Parade Marshalls George and Olive Brown with Fair Queen Janet Oilar, 1966.

and proud of the parade that he coordinates.

Everett could not have done it alone and he gives much praise to Lawrence Agee, Cecil Ray, Wayne Eldridge, Jim Albaugh and others that made the parade move smoothly. The Fall River Lions Club have taken over the job of controlling the traffic.

Everett also was involved with the Fire Department and was the chairman. Everett is now the administrator of Mayers Hospital and though he is not the parade chairman, he still does all he can to participate in the success of the fair.

Every year a grand marshall is honored at the fair parade. Chosen by the parade officials this person or persons may be young or old, but has accomplished the honor of being admired by the community.

A complete list of Grand Marshalls has not been kept in the past and therefore is incomplete. Those honorable citizens that do come to mind are Reuben Albaugh, Kenneth McArthur, Hack and Lil Lambert, Mike McCullough, Everett Beck, Tami Vestal, Ed and Pauline Bruce, June Vestal, Rhondo O'Conner, Assemblyman, Stan Statham; Pete Gerig and Aubrey Bieber, John and Rose Bartle, George and Olive Brown, Asa Doty, George Ingram, Norman Waggoner, George and Clover Corder, Assemblywoman, Pauline Davis; Senator, Ray Johnson; John Caton, Dr. Noecker, Mr. & Mrs. Dick Gallagher, Norman and Betty Taylor, Don and Rose Allison, Rose and David Schneider and the Fair Board, Willis Albaugh, Wally Hilliard, Catherine Ryan, Albert Albaugh and Willard Brown.

Greg Crum, 1969 Queen Sharon Telford and Duane Crum, ready for the parade.

Parade Marshalls Hack and Lil Lambert,
they hardly ever missed a fair
and always took part.
A legend in their time.

Tractors in the parade.

Reuben and Albert Albaugh, Parade Marshall, 1990.

The parade, 1983.

A little cowboy in the parade, 1989.

The Themes Contest

Each year for quite some time a theme has been used to highlight the fairs booths, flower borders, parade, activities and decorations.

A contest is held to have community members participate in the theme competition. A monetary award is given to the person who is lucky enough to have their theme chosen.

The Fair premium book cover depicts the theme and Alexis Ingram Johnson shares her artistic talent and has designed and drawn the cover for many of the programs.

The first theme contest was held and the first official theme was used at the 1960 Fair. That theme was "Through The Eyes of A Child".

Other themes through the years have been, Make Mine Country Style, Today, Wheels, 50 Golden Years, Show Me The 70's, Western Holiday, Our Country, Fair Time Is Fun Time, Happiness Is Going To The Inter-Mountain Fair, Country Living, Mountain Festival U.S.A., The Year Of The Horse, Our Precious Water, Western Fun In 81, Indian Summer, It's A Family Affair, There's A Song In The Air, An Old Fashioned Fair, Something Old Something New, Harvest Time, Calico Dreams n' Cowboy Things, Denim and Lace, Happy Birthday Lady Liberty, Ribbons and Bows and Cowboy Clothes and the 1995 fair had the theme that said it all, Showtime In The Mountains.

The decorating at the fair and the flower arrangements and borders depict the fair theme. Also the parade conveys the theme in some of their ways.

The Inter-Mountain Fair Logo, the McLaughlin Steam Tractor.

The Steam Tractor from the Past

The steam tractor is the logo synonymous with the Inter-Mountain Fair.

Built by the McLaughlin Manufacturing Company of West Berkeley, California this renowned piece of machinery has quite a history in the Inter-Mountain area. This tractor is only one of three ever manufactured of this style.

First bought by the John McArthur Company in 1908, its purpose was to pull the plows to farm the swamp land. The monstrous machine weighs 14 tons without the added water that the five hundred gallon tank holds. The tractor was delivered to Bartle by railroad and unloaded there and had to be driven to McArthur. Levi Lindsay drove the massive machine into the valley to the McArthur swamp.

Built for steam plowing, harvesting, lumbering, ore hauling and farming purposes it was thought to be the wonder of the ages. This high grade traction engine will supersede the draft animal. It has actual 60 Horse Power at the draw bar. Advertised that it will not do

Loading the Steam Tractor in Cayton Valley about 1952, (notice the water tank missing).

hauling as cheaply as can be done with horses. A ground speed of two and one half miles per hour hauling 40 to 50 ton. The massive machine makes a turn on a radius of forty feet.

The steam boiler burns two and one half cords of wood per day, and can plow four acres per hour pulling 40 plows cutting a 33 foot swath. Only an engineer, fireman and brake tender are needed to operate the enormous machine.

The original patent was applied for in 1903 and the reverse gear was patented in 1910.

For some reason enough steam could not be generated and the steam tractor was too heavy for the boggy ground and would sink so it was useless to the McArthurs. The Bosworth family of Cayton Valley bought the tractor and planned on using it to run a sawmill. Before they had their mill built, the Horr mill was put into production and Bosworths decided not to built one.

Levi Lindsay again climbed aboard the huge tractor to drive it to Cayton Valley for the Bosworths. Originally there were dual wheels to keep the heavy engine

from sinking into the ground while plowing. One rim came loose and they lost it near the airport. Levi then removed the other rim and left them both there. On Halloween one year some prankster boys rolled one of the wheels down the hill into the river. It landed in the river on the Knoch ranch and today when the river is low it can still be seen. The other rim was taken to Big Valley to be used as a cattle water trough.

The tractor set for many years not being used when the Bosworth family decided to send it to the fair as an artifact. In the early 1950's the tractor was loaded on a low boy truck and hauled to the fair. Sometime while it was at Cayton Valley the water tank was removed and used for a water truck. During a logging operation in that area it was shoved about with a caterpillar doing some damage to the gears. When the tractor arrived at the fair the water tank was offered to be put back on the tractor. Jim Norris took over the job of restoring this antique.

Today the McLaughlin Steam Tractor is the logo of the Inter-Mountain Fair and is an intriguing piece of history of the farming era in the Fall River area.

Moving the steam tractor to a new location at the fair grounds.